Tsietsi Jefrey Pilusa

Eco-fuel Briquettes

D1810108

Tsietsi Jefrey Pilusa

Eco-fuel Briquettes

Process Parameters and Conditions for Batch Production of Eco-Fuel Briquettes

LAP LAMBERT Academic Publishing

Impressum / Imprint
Bibliografische Information der Deutschen Nationalbibliothek: Die Deutsche Nationalbibliothek verzeichnet diese Publikation in der Deutschen Nationalbibliografie; detaillierte bibliografische Daten sind im Internet über http://dnb.d-nb.de abrufbar.
Alle in diesem Buch genannten Marken und Produktnamen unterliegen warenzeichen-, marken- oder patentrechtlichem Schutz bzw. sind Warenzeichen oder eingetragene Warenzeichen der jeweiligen Inhaber. Die Wiedergabe von Marken, Produktnamen, Gebrauchsnamen, Handelsnamen, Warenbezeichnungen u.s.w. in diesem Werk berechtigt auch ohne besondere Kennzeichnung nicht zu der Annahme, dass solche Namen im Sinne der Warenzeichen- und Markenschutzgesetzgebung als frei zu betrachten wären und daher von jedermann benutzt werden dürften.

Bibliographic information published by the Deutsche Nationalbibliothek: The Deutsche Nationalbibliothek lists this publication in the Deutsche Nationalbibliografie; detailed bibliographic data are available in the Internet at http://dnb.d-nb.de.
Any brand names and product names mentioned in this book are subject to trademark, brand or patent protection and are trademarks or registered trademarks of their respective holders. The use of brand names, product names, common names, trade names, product descriptions etc. even without a particular marking in this works is in no way to be construed to mean that such names may be regarded as unrestricted in respect of trademark and brand protection legislation and could thus be used by anyone.

Coverbild / Cover image: www.ingimage.com

Verlag / Publisher:
LAP LAMBERT Academic Publishing
ist ein Imprint der / is a trademark of
AV Akademikerverlag GmbH & Co. KG
Heinrich-Böcking-Str. 6-8, 66121 Saarbrücken, Deutschland / Germany
Email: info@lap-publishing.com

Herstellung: siehe letzte Seite /
Printed at: see last page
ISBN: 978-3-8454-2097-4

Zugl. / Approved by: South Africa, University of Johannesburg, Diss.,2010

Acknowledgements

The author acknowledges the National Research Foundation (NRF) of South Africa for financial assistance, without which the work as described in this dissertation would have not been possible.

Abstract

In this work, eco-fuel briquettes made from a mixture of 32% spent coffee grounds, 23% coal fines, 11% saw dust, 18% mielie husks, 10% waste paper and 6% paper pulp contaminated water, respectively were investigated. Various processing stages such as briquetting, drying, combustion and flue gas emissions were investigated in order to evaluate the socio-economic viability of the batch production of eco-fuel briquettes from biomass waste material. Each stage was studied independently in order to develop basic models that contained material and energy balances. A screw press briquetting machine was designed and fabricated as part of this work to be tested against the legacy foundation Porta press, and the Bikernmayer hand brick press. The compaction of the biomass waste material into briquettes follows the principle of physical interlocking of the fine particles within the plant fibres, natural material binding due to released cellulose content, as well as reduction in porosity, due to a simultaneous dewatering and compaction action. The processing variables such as cycle times and pressure were studied. The Bikernmayer press is preferred as it produced briquettes of higher bulk densities and lower moisture content as compared to the other presses. The drying was investigated in a laboratory scale convective dryer to establish typical convection parameters. A drying system that utilizes produced briquettes as a heating medium is proposed, and here drying will be effected over a refractory brick fireplace by means of convection and radiation. A basic model was set up to include radiation with the convection to predict a drying time of 4.8 hours. The combustion of briquettes was investigated using a POCA ceramic stove linked to the testo Portable Emission Analyzer System. This enabled an air-to-fuel ratio of 1.44 and a burning rate of 2g per minute to be established. The energy transfer efficiency for boiling a pot of water was found to be 85%. The gas emissions were found to be within the acceptable limits, as set out by OSHA. A standard initial economic evaluation was performed based on a briquette selling price of R2.26 per kilogram for the ease of accommodating the local market. The financial model for both Porta-press and screw press were not economically viable, as their running costs were greater than the gross project revenues. For the Bikernmayer conceptual model, with a total capital investment of R669, 981+ VAT (this includes one year operating cost) and a project life of five years, the gross profit margin is 44%, the profitability index is 5.33 and the internal Rate of return 31.44%. The net present value and return period are R676, 896 and 0.408 years respectively.

The customer profile as currently at hand is 17% of the selected area within 80 m radius from production site. The remaining 83% will be in need of energy as they become aware of the new product offering. The selling of the briquettes should be accompanied by an education process, to avoid the dangers of heating indoors. The principal driver for this project is socio economic development and it is being strengthened by Eskom's inability to provide sufficient energy. A secondary driver is the global drive to reduce emissions and fossil fuel usage; this technology does exactly this whilst diverting waste from landfill. In the Polokwane declaration (2008), it is stated that South Africa will have no calorific waste to landfill by 2014. Hence legislation will also provide a major part of the drive.

TABLE OF CONTENTS

CHAPTER 1: Project background

1.1 Previous study

A pre-feasibility study was conducted by the Phumani Paper in November 2006, in order to investigate the possibility of producing eco-fuel briquettes from waste paper and plant fibers such as mielie husks, river reed and sisal waste. At this stage, mixtures of the above raw material were prepared randomly by partial decomposition, and the resulting pulp was pressed by hand in a plastic cup. The briquettes were sun-dried and tested in the field by boiling one litre of water in a tin. During this study, a number of important factors were neglected, for example, briquetting pressure, moisture content, drying characteristics, net heating value, gas emissions, as well as basic heat transfer calculations. However, the study formed a good foundation in terms of development.

The study was later handed over to the chemical engineering department as a project to design a complete process of producing eco-fuel briquettes. The proposed briquettes were to be produced using the Porta press equipment from the Legacy foundation. The primary objective was to propose a suitable production method using mielie husks, cardboard and river reed as raw materials. There were a suspected number of complications in preparing the raw material, for instance cooking dry river reed and mielie husks in soda ash for hours before it can be put in the sun (in sealed black plastic bags) for weeks to allow further decomposition. A number of tests were conducted on the following: material preparation, blending, pressing, drying as well as quality testing of the final product. This was done in order to determine the suitable processing conditions of the proposed briquette type. The main disadvantage of this method was energy requirements and time. In addition, the final product took longer to dry in an oven due to high moisture content, with a resulting lower dry bulk density and shorter burning times.

With a view to improving the briquetting scene of batch production of eco-fuel briquettes in the townships of Gauteng, the Gauteng Department of Agriculture has financed this specific project through Phumani Paper in order to develop an eco-friendly batch production system, and also to engage communities by teaching them various methods of utilising domestic waste. The use of available natural resources has become a viable and economical option to energy savings. People living in the townships find it more convenient to use all possible cost effective energy sources for heating and cooking. Further research has indicated that a number of industries in are disposing large quantities of waste material on a daily basis which could potential be a raw material for the eco-fuel briquettes as stipulated in the Enviroserv waste disposal database.

1.2 The fuel briquettes

Generally, a fuel briquette is made from any carbon containing compound, which is combustible and produce less toxic gasses. Theoretically, more than 80% of organic compounds contain carbon; however the issue of combustibility and emissions becomes a determining factor.

The proposed research idea is focusing on various raw materials such as: spent coffee grounds, coal fines, grass clippings, wood chips and other biodegradable organic material. The proposed fuel briquette is basically a round disc made of slightly decomposed and compressed spent coffee beans, grass and coal fines,

2

when burned it produces heat energy. The disc has a hole in the centre to allow the flame to breath and burn. The proposed product will serve as an alternative energy source to the available energy sources. As a matter of environmental consideration the combustion gasses produced by the briquette will be tested to ensure that it conforms to the minimum requirements for gas emissions.

Due to rising costs of electrical energy, heat for cooking food and heating in most townships is obtained by paraffin heating and the burning of coal and wood. The recent market survey has indicated that communities in the townships uses large quantities of coal during winter for cooking, and to keep their homes warm. The survey has also indicated that approximately 85% of the people living in the townships use coal stoves to burn coal in winter for heating and cooking. It costs seventy five rand (R75.00) to buy a 25kg bag of coal, which last for a week, and is equivalent to R280.00 per month, in addition to electricity used to operate other home appliances. The general response of the survey has shown a potential market demand for this product, and that serves as a primary motivation to carry out this study. Results shown in table 1.1 show the comparisons of various alternative fuels available.

Type	Electricity	Coal	Wood	Paraffin	LP Gas	Cobble stones	Eco-fuel briquettes
Rate	R0.45/kW.h	R2.60/kg	R2.80/kg	R13.50/kg	R15/kg	R30.55/kg	R2.26/kg
Energy content	3.6MJ/kWh	24.3MJ/kg	16MJ/kg	42MJ/kg	49.3MJ/kg	19.7MJ/kg	18.9MJ/kg
Fuel cost	12.5Cents/MJ	10.7Cents/MJ	17.5 Cents/MJ	32 Cents/MJ	30.4 Cents/MJ	155Cents/MJ	11.9Cents/MJ

Table 1.1: Alternative fuels comparison (Data obtained in the form of questioners survey conducted as part of this research)

1.3 Project motivation

Phumani Paper, a non-Government organisation in Doornfontein has successfully developed several small business enterprises in hand papermaking and craft production across South Africa. The business units produce paper with area-specific fibres and invasive plant species. A key challenge experienced by most rural and urban business units is to sustain and increase their income base through extensive product development. Phumani Paper is also actively involved in assisting community-based enterprises to diversify their income base by introducing new and innovative products. This requires, in some cases, focusing on innovative products, using existing skills and capacity. The process utilised in hand papermaking of the breaking down of plant fibres is similar to that of the method used in the production of eco-fuel briquettes. Therefore, it would be a viable option to implement the production of fuel briquette into those business units as an added benefit to income generation. Due to its location and supply of raw materials, small paper-making enterprises already have some understanding of eco-friendly conservation techniques, which will further be enhanced with the implementation of the project.

1.4 Research Feasibility

The project has already been set up with all the necessary equipment and laboratory facilities. It is believed that with adequate effort and commitment, the project can be completed successfully within the given time constraints. The project will be conducted in cooperation with Phumani Paper, and it is believed to have potential positive impact, since a key challenge experienced by most rural and urban business units is to sustain and increase their income base through extensive product development. Phumani Paper is also actively involved in assisting community-based enterprises to diversify their income base by introducing new and innovative products. This requires, in some cases, focusing on innovative products, as well as using existing skills and capacity. Industrial waste can be utilised in the manufacture of briquettes, thereby preserving the environment through the utilisation of unwanted plant matter, as opposed to depleting the natural forest.

The briquette is an energy efficient alternative and is especially attractive to areas with little or no electricity, It is a relatively cheap source of energy, the manufacture of fuel briquette can aid in job creation. The promotion of eco-fuel briquettes can boost the local economy (F. Meintjies, personal communication, June 7, 2007).

1.5 Research Design and Methodology

The research objective determines the selection of appropriate research method and reliable data collection techniques. The proposed research methodology for this project is designed in such a way that all objectives will be fully achieved within the given time constraints and availability of resources. The proposed eco-fuel briquettes will be produced on a laboratory scale and all the necessary parameters will be modelled in order to give an indication of how bulk production will be carried out.

The processing of fuel briquette making materials is the key step in fuel briquette work. It involves chopping, wetting and resting of the materials until the materials are ready to be pressed. The principle on which the fuel briquettes are bound

5

together is basically a mechanical process, in which naturally occurring residues are first chopped, then softened through partial decomposition. They are then blended with other agro residues or pre-processed residues such as sawdust or coal fines. This blending process causes the fibres to be randomly redistributed throughout the mass, and to entangle and interlock into a solid mass as the mass is compressed and dewatered. There is no chemical binder used in this fuel briquette making process. Fuel briquette making theory is based on the tendency of natural fibres to interlock when combined with other by-products of agro processing. The objective is to create a solid compact material without the introduction of any glue, resin or other artificial binders.

1.6 Availability and collection of raw material

The spent coffee beans will be obtained from the National Brands limited. The use of coal fines as an additive to the coffee beans will be considered, and this is due to the fact that coal is a natural fossil fuel and has a high net heating value (Jameson, 1988). Coal fines are readily available in most townships around the east rand, they are usually dumped in open spaces as an end waste product of the coal packaging process and can thus be obtained free of charge. Mielie husks are obtainable as an agricultural waste product from the farms just before winter season. Decomposed mielie husks and waste paper can be added to the feed mixture to enhance briquette binding and stability. Waste paper is available in academic institutions and can be obtained free of charge. The use of a chemical binder will not be considered due to the operational cost and environmental considerations. The use of partially decomposed waste paper will be considered as a binding agent instead of chemical binders.

1.7 Research objectives

The main objective of this is to pave the way for domestic batch production of environmentally friendly eco-fuel briquettes from agro waste and various natural organic waste materials. The research project focuses on the investigation of suitable production process parameters and conditions. The proposed study outcomes include developing systematic model to be used in monitoring process conditions for a desired briquette production to be used as an alternative energy source. It will also cover aspects such as environmental conservation through conversion of industrial and agricultural waste material such as spent coffee beans, coal fines, wood chips, grass, mielie husks and other various plant fibres into useful products. The research output is aimed at engaging and developing poor communities through skills and knowledge transfer, furthermore the end product will aid with the supply of alternative energy source and job creation.

Specific Objectives:

- Design and fabricate a briquetting machine.
- Make briquettes from the proposed biomass mixture.
- Measure process parameters such as mixing times, cycle times, moisture content, drying time, energy utilisation and combustion gasses quality.
- Develop a basic economic model using the measured parameters as inputs in order to evaluate the economic viability of the process.

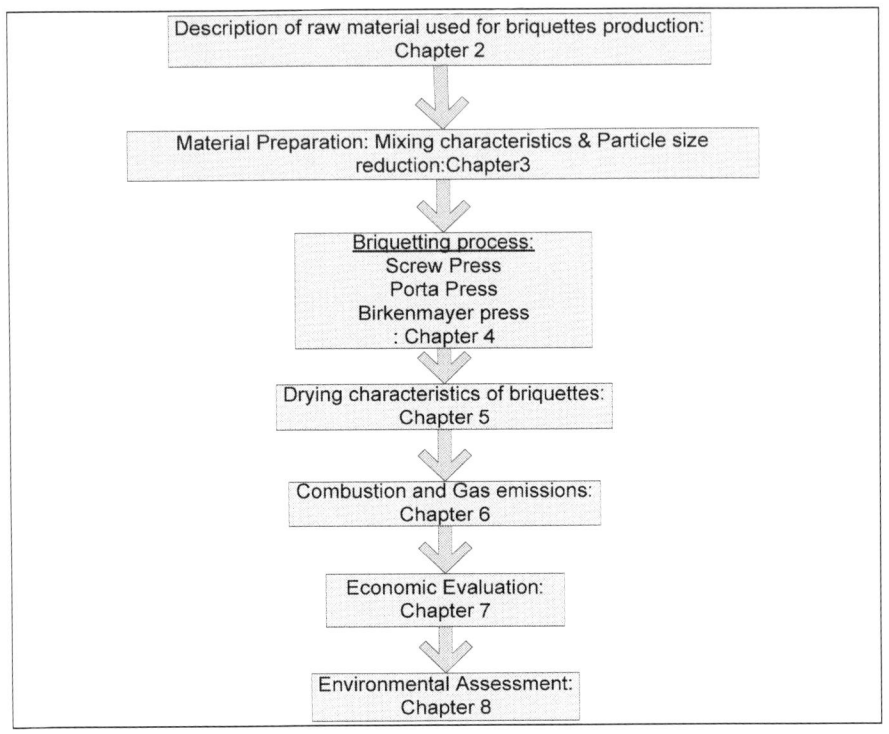

Figure 1.1: Structure of the research project

Figure 1.1 shows the structure of this dissertation; raw material preparation is investigated followed by the briquetting process. Once the briquettes are made, they are dried while investigating their drying characteristics. The briquettes burning profile and gas emission analysis is conducted and the data obtained is fed into the basic economic model, whereby the economic feasibility of the project is evaluated. The project feasibility is further justified by assessing the environmental friendliness of the emissions resulting from burning eco-fuel briquettes. In each chapter, there is a section on the background, method, experimental work conducted followed by the results, discussions and conclusions for the chapter.

8

CHAPTER 2: Introduction

2.1 History of fuel briquettes

Historically, biomass briquetting technology has been developed in two distinct directions. Europe and the United States has pursued and perfected the reciprocating ram-piston press, while Japan has independently invented and developed the screw press technology (Granada, Lopez Gonzelez, Miguez & Moran, 2002). Although both technologies have their merits and demerits, it is universally accepted that the screw pressed briquettes are far superior to the ram pressed solid briquettes in terms of their storability and combustibility (Debdoubi, El amarti & Colacio, 2005). Japanese machines are now being manufactured in Europe, under a licensing agreement but no information has been reported about the manufacturing of European machines in Japan. Worldwide, both technologies are being used for briquetting of sawdust and locally available agro-residues. Although the importance of biomass briquettes as substitute fuel for wood, coal and lignite is well recognized, the numerous failures of briquetting machines in almost all developing countries have inhibited their extensive exploitation (Yaman, Sahan, Haykiri-acma, Sesen and Kucukbayrak, 2000).

2.1.1 Biomass densification

Biomass densification, which is also known as briquetting of sawdust and other agro residues, has been practiced for many years in several countries. Screw extrusion briquetting technology was invented and developed in Japan in 1945. As of April 1969, there were 638 plants in Japan engaged in manufacturing sawdust briquettes (Wamukonya & Jenkins, 1994). The fact that the production of briquettes quadrupled from 1964 to 1969 in Japan speaks for the success of this technology. At present, two main high-pressure technologies: ram or piston press and screw extrusion machines, are used for briquetting. While the briquettes produced by a piston press are completely solid, screw press briquettes on the other hand have a concentric hole which gives better combustion characteristics due to a larger specific area. The screw press briquettes are also homogeneous and do not disintegrate easily. Having a high combustion rate, these can substitute for coal in most applications and in boilers (Debdoubi *et al*, 2005).

Briquettes can be produced with a density of 1200kg/m³ from loose biomass of bulk density 100 to 200kg / m³ (Ndiema, Manga & Ruttoh, 2001). These can be burnt clean and are therefore eco-friendly arid also those advantages that are associated with the use of biomass are present in the briquettes. Biomass is matter usually considered as waste. Some of it exits as dead trees, tree branches, yard clippings, left-over crops, wood chips, bark and sawdust from lumber mills. It can even include food waste and livestock manure (Yaman et al, 2000).

2.1.2 Benefits of Using Biomass

Using biomass can help reduce global warming compared to a fossil fuel-powered plant. Plants use carbon dioxide (CO_2) when they grow. CO_2 stored in the plant is released when the plant material is burned or decays. By replanting the crops, the new plants can use the CO_2 produced by the burned plants. So using biomass and replanting helps close the carbon dioxide cycle. However, if the crops are not replanted, then biomass can emit carbon dioxide that will contribute toward global warming (Demirbas & Sahin, 1997).

Biomass can be used for fuels, power production, and products that would otherwise be made from fossil fuels. In such scenarios, biomass can provide an array of benefits. For example, the use of biomass energy has the potential to greatly reduce greenhouse gas emissions. Burning biomass releases about the same amount of carbon dioxide as burning fossil fuels. However, fossil fuels release carbon dioxide captured by photosynthesis over millions of years, essentially new greenhouse gas. Biomass, on the other hand, releases carbon dioxide that is largely balanced by the carbon dioxide captured in its own growth, depending how much energy was used to grow, harvest, and process the fuel. The use of biomass can reduce dependence on foreign oil because biofuels are the only renewable liquid transportation fuels available.

2.2 The art of fuel briquettes production

The proposed fuel briquette is basically a round disc made of slightly decomposed and compressed plant matter. The disc has a hole in the centre to allow the flame to breath and burn. The briquette is an alternative source of energy for cooking and heating and is especially important in those areas with little or no electricity. With the rising costs of energy, it provides an affordable and environmentally friendly alternative energy source for the poor.

The production of the fuel briquette has been researched and implemented in Unites States of America several African such as Uganda, Rwanda, Kenya and Malawi and Malaysia. The methodology behind the briquettes has enormous potential for people utilising wood as a fuel for cooking and heating. By using agricultural waste or invasive vegetation as the raw material in the production of the briquettes valuable wood lots and biodiversity can be protected (F. Meintjies, personal communication, June 7, 2007).

Figure 2.1: Corn stover briquette (Source: Mani, Tabil & Sokhansanj, 2006).

Figure 2.1 shows briquettes made from corn stover, a major field of crop residues in the United States of America, comprising roughly of 75% of total agricultural residues with an average bulk density of 42 kg/m^3. These briquettes are produced using a hydraulic piston and cylinder press at pressures of 5-15 MPa. Comparison details with other briquettes are listed in table 2.1.

Figure 2.2: Cobb Cobblestone briquettes (February 18, 2010 http://www.cobbglobal.com/cobblestone.asp).

The Cobb Cobblestones briquettes are produced from coconut shells, they provide over 2 hours burning time with each briquette weighing 420g. These briquettes are very compact compared to others and they are made under extremely high pressure with the aid of a chemical binder.

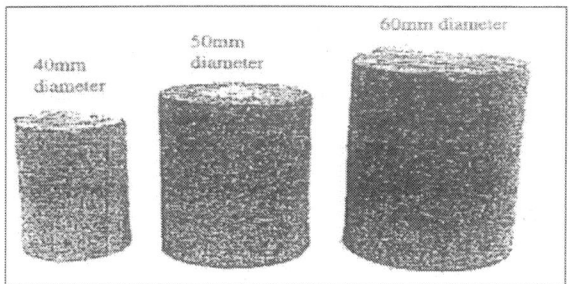

Figure 2.3: Palm fibre and Shell briquettes (Husain, Zainac & Abdullah, 2002) These briquettes are made from palm shell and residue in Malaysia; they are available in 40mm, 50mm and 60mm diameters. They are also made using a hydraulic press at pressures of 5-13.5 MPa but has a higher density (1200kg/m^3) when compared with other briquettes.

Figure 2.4: The eco-fuel briquettes, made from spent coffee beans, coal fines, waste paper, saw dust and mielie husks (Picture taken at Thuthukani, Tsakane).

The eco-fuel briquettes are made using a very low pressure hand operated Porta press or screw press. All briquettes have an outer diameter of 100mm, inner diameter of 35mm and 50mm long. There is no use of a chemical binder; the material undergoes natural binding by interlocking themselves between partially decomposed plant fibres. They are not as compact as compact as cobb-cobblestones but yet have longer burning times and higher gross calorific values.

Name	Material	Press type	Pressure	Moisture	Density	Calorific Value	Diameter	Length	burning rate
Eco-fuel briquettes	Coffee beans, coal fines, paper, saw dust & mielie husks	Hand press	0.1-2.2 MPa	11%	213-732 kg/m³	18.9 MJ/kg	100mm	50mm	2 g/ min
Corn stover briquettes	Corn stover	Hydraulic press	5-15 MPa	10-15%	650-650 kg/m³	16.1 MJ/kg	32mm	25mm	Not measured
Cobb-Cobblestones	Coconut shell	Hydraulic press	Not measured	9%	1112 kg/m³	14.8 MJ/kg	135mm	30mm	3.5 g/min
Palm briquettes	Palm fibres and shells	Hydraulic press	5-13.5 MPa	12%	1200 kg/m³	16.4 MJ/kg	40-60mm	53-80mm	Not measured

Table 2.1: Comparison of fuel briquettes

The data shown in table 2.1 proves that the eco-fuel briquettes are comparable with other briquettes made in the past irrespective of their low compaction pressures. They are cheaper to produce due to the low briquetting pressure but yet provide highest gross calorific value with longer burning times.

2.2.1 The need for the production of fuel briquettes in South Africa

The use of eco-fuel briquettes as an alternative source of energy has several benefits in the South African context:

- Agro waste can be utilised in the manufacture of briquettes thereby preserving the environment through the utilisation of unwanted plant matter, as opposed to the depletion of natural forest;
- The briquette is an energy efficient alternative and is especially attractive to areas with little or no electricity;
- The manufacture of eco-fuel briquettes can aid in job creation;
- Eco-fuel briquettes are a relatively cheap source of energy that can ultimately provide jobs that are in line with the Government's plan for the sustainable use of natural resources.

2.3 The process of briquettes production

The processing of fuel briquette materials is the key step in fuel briquette work. It involves chopping and wetting of the materials until the materials are ready to be pressed. The principle on which the eco-fuel briquettes are bound together is basically a mechanical process, in which naturally occurring residues are first chopped, then softened through partial decomposition. They are then blended with other agro residues or pre-processed residues such as sawdust or coal fines. This blending process causes the fibres to be randomly redistributed throughout the mass, and to entangle and interlock into a solid mass as the mass is compressed and dewatered. There is no chemical binder used in this fuel briquette making process. Fuel briquette making theory is based on the tendency of natural fibres to interlock when combined with other by-products of agro processing. The objective is to create a solid compact material without the introduction of any artificial binders.

2.3.1 Decomposing the materials

As with composting, natural material decomposes quickly when it is sealed, moist and out of the wind. Some material will require only a few days; others will require several weeks depending upon the local climate and the material used. The materials should be left to decompose until they become soggy and warm to the touch. Mixing may be used for further breaking up the materials to assure uniformity in the stage of decomposition.

The time required for decomposition depends upon the season and the material type. Generally allow two weeks in a hot/dry season and up to six weeks in a cold season, with times for decomposition during the rainy seasons, falling somewhere in between. In the case where the material cannot be used immediately, they can be stockpiled with no cover to stop, or at least dramatically slow down, further decomposition. Once interrupted and left in dry, open, aerobic conditions, the prepared material can be stored until it is needed. Leaving it in dark humid conditions however, will further soften and destroy fiber strength resulting in weak eco-fuel briquettes. The use of pre-processed residues such as sawdust, charcoal fines or spent coffee beans can reduce the above suggested agro-residue processing volume by 40% as these pre-processed wastes are simply blended into the agro-residues when the latter are ready. The fibers should not be left to decompose or burn up beyond this point. If left too long, they will begin to break down like compost, generating basically, rich humus, which does not burn very well.

2.3.2 Mixing of raw materials

After decomposition, once the materials have been suitably decomposed they are ready for blending. They are diluted in water and mixed into a coarse paste or sludge consistency so that they can be easily added to the mould. This mixing also ensures that the fibers are randomly distributed throughout the mass around them. At this stage the material is ready for pressing.

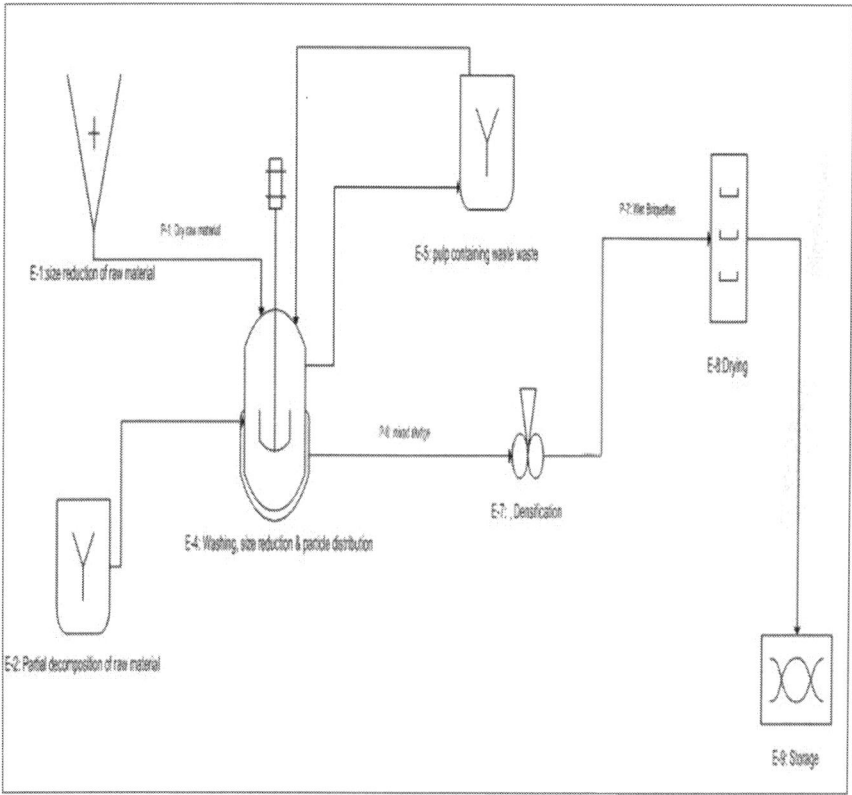

Figure 2.5: Schematic Diagram of fuel briquette processing setup (drawn using Microsoft Visio, ver. 2003).

The above figure shows the necessary steps required for eco-fuel briquettes production. The first step is size reduction and screening of dry raw material followed by soaking of the biomass material to allow slight decomposition this is done by soaking the biomass in pulp containing waste water from the Phumani Paper archive mill. The material is fed into the mixer at the right proportion and mixed until uniformity is achieved. The resulting blend is densified into briquettes using a screw press, Porta press and Bikernmayer press. The final unit operation stage in this process is drying followed by storage.

Raw Material blend	Sample Mass(kg)	% (w/w)
Spent coffee ground	4.3	31.9
Finely ground coal fines 80%<500µm	3.1	23.0
Saw dust	1.5	11.1
Mielie husk/grass clippings/cow dung	2.4	17.7
Shredded Paper(news print/card board)	1.4	10.3
Pulp containing water	0.8	5.9
Total material	13.5	100

Table 2.2: Proposed recipe for eco-fuel briquettes production.

Table 2.2 shows how the proposed briquette is made up in terms of raw material composition. The primary ingredient of the briquettes is spent coffee beans from the National Brands Limited constituting 31.9 mass percent of the blend. The secondary raw material is 23 mass % coal fines, and other biomass material as indicated in table 2.2. The use of pulp contaminated water from the Phumani Paper production is unique is such a way that fibres already contained in the water is necessary for binding during the briquetting process. The use of this water is also an environmental responsible method of managing the effluent from the archive paper production.

Ultimate analysis	Carbon	Hydrogen	Oxygen	Nitrogen	Sulphur	Calorific Value
Spent coffee ground	33.50%	4.20%	20.37%	1.14%	0.03%	14.5 MJ/kg
Coal fines	51.60%	6.30%	24.80%	0.67%	1.23%	21.3 MJ/kg
Sawdust	42.38%	5.27%	42.41%	0.14%	0.01%	16.7 MJ/kg
Mielie husks	42.10%	7.20%	43.60%	1.20%	0.21%	18.2 MJ/kg
Waste paper	43.40%	5.80%	44.30%	0.25%	0.34%	13.8 MJ/kg

Proximate Analysis	Ash	Moisture	Fixed Carbon	Volatile Matter	Density
Spent coffee ground	0.66%	40.00%	19.60%	39.74%	427 kg/m^3
Coal fines	9.70%	5.70%	55.50%	29.10%	560 kg/m^3
Sawdust	1.55%	8.25%	14.04%	76.16%	290 kg/m^3
Mielie husks	1.58%	4.11%	18.40%	77.49%	104 kg/m^3
Waste paper	0.80%	7.70%	11.79%	82.30%	210 kg/m^3

Table2.3: Chemical properties of the raw material used to make eco-fuel briquettes as obtained from source.

The above chemical analysis was measured using an Inductively Coupled Plasma mass spectrometry (ICP-Model- Optima 2100DV) at Enviroserv Reitfontein laboratory. Sample of each raw material was sent to the laboratory for ultimate and proximate elemental analysis.

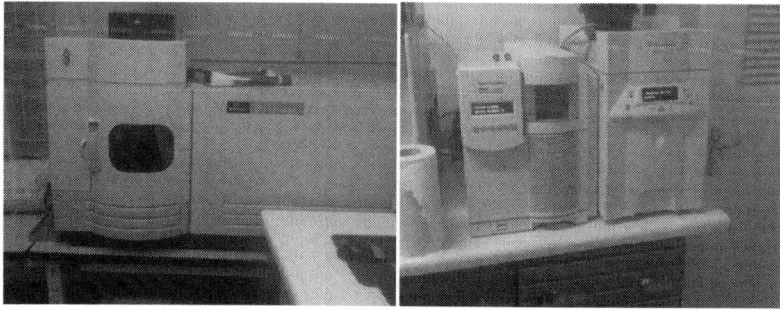

Figure 2.6: Inductively coupled plasma mass spectrometry and gas chromatography (Model: Optima 2100DV) used for ultimate and proximate analysis of the raw material (Picture taken at EnviroServ Reitfontein laboratory).

@-No / last=338	waste or fuel	ref.	remarks	wet basis water w kg/kg wet	proximate analysis (dry basis) fixed carbon kg/kg dry	volatile matter kg/kg dry	ultimate analysis (massfraction, sum = 1) (dry mass basis) ash,inert a kg/kg dry	carbon c kg/kg dry	hydrogen h kg/kg dry	oxigen o kg/kg dry	nitrogen n kg/kg dry	sulfur s kg/kg dry	chlorine cl kg/kg dry	lower heating value calculated Hu MJ/kg dry	upper heating value calculated Ho MJ/kg dry
26	carbon (pure) Kohlenstoff (rein)	/1/		0			0	1	0	0	0	0	0	34.91	34.91
14	anthracite Anthrazit (Ruhr)	/2/		0.05			0.03158	0.89474	0.03158	0.02105	0.01053	0.01053		34.07	34.76
55	bituminous coal (Goettelborn, D)	/11/	*55	0.089	0.573	0.32931	0.09769	0.73798	0.04896	0.09133	0.01405	0.00801	0.00199	29.37	30.44
35	lignite Braunkohle, roh (Rheinland)	/1/		0.5			0.1	0.6	0.06	0.2	0.02	0.02		24.60	25.91
37	peat (air-dry) Torf, lufttrocken	/1/		0.25			0.06667	0.50667	0.05333	0.34667	0.01333	0.01333		19.20	20.36
105	wood unbelasteled Holz	/25/		0.447			0.00579	0.49801	0.0642	0.43146	0.00011	0.00011	0.0004	19.08	20.48
106	waste wood Altholz	/25/	*106	0.05			0.16905	0.414	0.04905	0.36074	0.00305	0.00011	0.004	15.07	16.14
138	corn	/31/	*138	0.139	0.18049	0.80372	0.0150	0.42001	0.07236	0.47631	0.01254	0.00209		16.09	18.26
1	municipal solid waste Hausmüll	/2/		0.241			0.2556	0.37418	0.05138	0.29908	0.01186	0.00132	0.00659	14.36	15.48
260	mixed paper	/46/,/48/	*260	0.1024	0.094	0.846	0.06	0.4341	0.0582	0.4432	0.0025	0.002		16.05	17.32
7	plastic wastes Kunststoffabfälle	/2/		0.15			0.10118	0.66235	0.09176	0.09529	0.01059	0.00353	0.03529	30.75	32.76
54	scrap tyre	/10/		0.0099	0.17614	0.63115	0.19271	0.71498	0.06878	0.00707	0.09606	0.0104		31.18	32.68
259	vegetable food waste (1972)	/46/,/48/	*259	0.7829	0.1635	0.7877	0.0488	0.4906	0.0662	0.3756	0.0168	0.002		19.49	20.94
10	garden wastes Gartenabfälle	/2/		0.45			0.18364	0.42364	0.05273	0.31818	0.01636	0.00364	0.00182	16.19	17.34
4	sewage sludge Klärschlamm entw.	/2/		0.742			0.36850	0.30233	0.04651	0.24806	0.03191	0.0155		11.81	12.83
151	dairy free-stall manure	/35/	*151 #2	0.703	0.07071	0.3064	0.6229					0.0001			
24	light heating oil Heizöl EL	/2/	*24	0.004			0	0.86345	0.13052	0.00402	0	0.00201		42.65	45.50
155	earth's crust Erdkruste	/36/	*195 #1	0			0.0002	0.0014	0.466	0.00002	0.00026	0.00013		

Figure 2.7: Ultimate and proximate analysis of waste raw material and fuels (Wochele, 2003).

The test results presented in table 2.3 appear to be reasonable when compared with the ones suggested by Wochele (2003) in table 2.7.

2.4 References

Debdoubi A., El amarti A. & Colacio E., 2005. Production of fuel briquettes from esparto partially pyrolyzed. *Energy Conversion and Management*, 46: 1877-1884.

Demirbas A. & Sahin A., 1997. Evaluation of biomass residue: Briquetting waste paper and wheat straw mixtures. *Fuel Processing Technology*, 55: 175-183.

Granada E., Lopez Gonzelez L.M., Miguez J.L. & Moran J., 2002. Fuel lignocellulosic briquettes die design and product study. *Renewable Energy*, 27: 561-573.

Husain, Z., Zainac Z. & Abdullah Z. 2002. Briquetting of palm fibre and shell from the processing of palm nuts to palm oil. *Biomass and Energy*, 22: 505-509.

Mani, S., Tabil L.G. & Sokhansanj S. 2006. Specific energy requirement for compacting corn stover. *Bioresource Technology*, 97: 1420-1426.

Ndiema, Manga & Ruttoh, 2001. Influence of die pressure on relaxation characteristics of briquetted biomass. *Energy Conversion and Management*, 43: 2157–2161.

Wamukonya L. & Jenkins B., 1994. Durability and relaxation of sawdust and wheat straw briquettes as possible fuels for Kenya. *Biomass and Bioenergy*, 8: 175-179.

Wochele, J. 2003. Composition and analysis of waste raw material and fuels. *A Compilation of Elemental Analysis Database*, 21: 17-21.

Yaman S., Sahan M., Haykiri-acma H., Sesen K. and Kucukbayrak S. 2000. Production of fuel briquettes from olive refuse and paper mill waste. *Fuel Processing Technology*, 68: 23-31.

CHAPTER 3: Material preparation

3.1 Background

Raw materials such as coal fines, mielie husks and paper often obtained in sizes that are too large to be used for the purpose of briquetting and, therefore, they must be reduced in size. Other raw material such as spent coffee beans, saw dust and grass clippings do not require size reduction as they are obtained in a particle size suitable enough for the briquetting process. This size-reduction operation can be divided into two major categories, depending on whether the material is a solid or a liquid. If it is solid, the operations are called grinding and cutting, if it is liquid, emulsification or atomisation. All depending on the reaction to shearing forces within solids and liquids (Alter, 1980).

Grinding and cutting reduce the size of solid materials by mechanical action, dividing them into smaller particles. In the grinding process, materials are reduced in size by fracturing them. The mechanism of fracture is not fully understood, but in the process, the material is stressed by the action of mechanical moving parts in the grinding machine and initially the stress is absorbed internally by the material as strain energy. When the local strain energy exceeds a critical level, which is a function of the material, fracture occurs along lines of weakness and the stored energy is released. Some of the energy is taken up in the creation of new surface, but the greater part of it is dissipated as heat. Time also plays a part in the fracturing process and it appears that material will fracture at lower stress concentrations if these can be maintained for longer periods. Grinding is, therefore, achieved by mechanical stress followed by rupture and the energy required depends upon the hardness of the material and also upon the tendency of the material to crack - its friability (Jackson, 1974).

Figure 3.1: Wet spent coffee beans as obtained from the National brands limited.

The spent coffee beans are rich in oils, water and cellulose and form a primary ingredient of the eco-fuel briquettes. They do not require any preparation in terms of size reduction as they break down easily when they are squeezed or agitated in a high speed mixer. Unlike waste paper and mielie husks, spent coffee beans do not require any cutting and wetting to free their fibres for the ease of mixing and biding. Spent coffee beans have a water saving advantage due to the fact that they obtained as a bi product of hot water soluble coffee extraction process with approximately 40% moisture content. No water is required for soaking and they are ready to be used from source.

Figure 3.2: Coal fines as obtained from an informal dump site in Etwatwa.

Coal fines are the secondary raw material of the eco-fuel briquettes due to their high calorific value and water resistance. However they are obtained in various particles sizes. One of the main disadvantages of coal fine is their binding and mixing characteristics with other material. Unlike other raw material used in this research for eco-fuel briquettes production, coal fines require intensive screening and crushing. For experimental purposes, a laboratory scale rod mill was used in this research for coal fines size reduction. Coal is considered soft to medium material in terms of Mohs scale of hardness(1-3) and has low Bond work index of 12.5 -13(kW.hr/short tonne) as indicated in table 3.1, which is equivalent to crushing energy requirements of 51.6 kilojoules per kilogram of coal crushed (Wills and Napier-Munn, 2006).

Mineral	Work Index (kW.hr/s-ton)	Mineral	Work Index (kW.hr/s-ton)
Barite	4.7 - 6.9	Glass	3.4
Basalt	17.0 - 22.5	Granite	15.1 - 16
Cement clinker	14.8	Limestone	9.0 - 12.8
Coal	12.5 - 13.0	Mica	148
Dolomite	9.0 - 12.4	Quartz	13.6 - 14.1
Feldspar	12.8	Quartzite	9.6

Table 3.1: Bond Work Index for various minerals (source: Wills and Napier-Munn, 2006).

The particle size distribution may be expressed as a range analysis, in which the amount in each size range is listed in order. It may also be presented in cumulative form, in which the total of all sizes retained or passed by a single notional sieve is given for a range of sizes. Range analysis is suitable when a particular ideal mid-range particle size is being sought, while cumulative analysis is used where the amount of under-size or over-size must be controlled (Teeuwen, 1980).

Methods may be simply shaking of the sample in sieves until the amount retained becomes more or less constant. Alternatively, the sample may be washed through with a non-reacting liquid (usually water) or blown through with an air current. The most obvious disadvantage is that the smallest practical sieve size is 20-40 μm, and many PSDs are concerned with much smaller sizes than this. A 20 μm sieve is exceedingly fragile, and it is very difficult to get material to pass through it.

Figure 3.3: Laboratory vibrating sieves used in this research for coal fines screening & particle size distribution

To find the percent of aggregate passing through each sieve, first find the percent retained in each sieve. To do so, the following equation is used,

$$\%\text{Retained} = \left[\frac{W_{sieve}}{W_{Total}}\right] \times 100\%\dots\dots\dots\dots\dots\dots\dots\dots\dots\dots\dots\dots 3.1$$

Where W_{Sieve} is the weight of aggregate in the sieve and W_{Total} is the total weight of the aggregate. The next step is to find the cumulative percent of aggregate retained in each sieve. To do so, add up the total amount of aggregate that is retained in each sieve and the amount in the previous sieves. The cumulative percent passing of the aggregate is found by subtracting the percent retained from 100%.

%Cumulative Passing = 100% - %Cumulative Retained

Figure 3.4: Grass clippings, mielie husks and shredded paper soaked in pulp contaminated water from Phumani Archive paper mill.

The mielie husks and paper are shredded and soaked in water with other biomass such as grass clippings and saw dust. This material is allowed to re-pulp and undergo a bit of anaerobic decomposition to improve their binding ability. For experimental purposes, the office waste paper was shredded using an office paper shredder and the cardboard was cut in smaller pieces using scissors. The dry mielie husks were also cut in to smaller pieces using scissors.

Decomposition of plant matter occurs in many stages. It begins with leaching by water; the most easily lost and soluble carbon compounds are liberated in this process. Another early process is physical breakup or fragmentation of the plant material into smaller bits, which have greater surface area for microbial colonization and attack. In smaller dead plants, this process is largely carried out by the soil invertebrate fauna, whereas in the larger plants, primarily parasitic life-forms such as insects and fungi play a major breakdown role and are not assisted by numerous detritivore species. Following this, the plant consisting of cellulose, hemicelluloses, microbial products, and lignin undergoes chemical alteration by microbes. Different types of compounds decompose at different rates (*Barnes, 1985*).

3.2 Theoretical basis

3.2.1 Energy requirements for coal grinding

Grinding is a very inefficient process and it is important to use energy as efficiently as possible. Unfortunately, it is not easy to calculate the minimum energy required for a given reduction process, but some theories have been advanced which are useful. These theories depend upon the basic assumption that the energy required to produce a change dL in a particle of a typical size dimension L is a simple power function of L (Barnes, 1985)

$$\frac{dE}{dL} = KL^n \dots\dots\dots 3.2$$

- dE is the differential energy required, dL is the change in a typical dimension, L is the magnitude of a typical length dimension and K, n, are constants.
- K_R is called **Rittinger's constant**, and integrate the resulting form of equation.

$$\int dE = K_R \int_{L_1}^{L_2} L^{-2} dL$$

$$\therefore E = K_k \left[\left(\frac{1}{L_2} \right) - \left(\frac{1}{L_1} \right) \right] \dots\dots\dots 3.3$$

As suggested by Bond (1952), the total work required for crushing is defined as follows:

$$E = W_i\left[\left(\frac{10}{\sqrt{x_2}}\right) - \left(\frac{10}{\sqrt{x_1}}\right)\right] \dots\dots\dots\dots\dots\dots\dots\dots\dots\dots\dots\dots\dots\dots\dots\dots\dots3.4$$

W_i : Bond work index(kW.h/s - ton)

x_1 : Initial particle size(m)

x_2 : Final particle size(m)

The index is defined as the energy required for crushing material from infinite particle size down to 100μm.

3.2.2 Rates of Mixing

Once a suitable measure of mixing has been found, it becomes possible to discuss rates of accomplishing mixing. It has been assumed that the mixing index is to be such that the rate of mixing at any time, under constant working conditions such as in a well-designed mixer working at constant speed, is to be proportional to the extent of mixing remaining to be done at that time. That is,

$$\frac{dM}{dt} = K[(1-(M))] \dots\dots\dots\dots\dots\dots\dots\dots\dots\dots\dots\dots\dots\dots\dots\dots\dots\dots\dots\dots3.5$$

(M) is the mixing index and K is a constant, and on integrating from $t = 0$ to $t = t$ during which (M) goes from 0 to (M),

$$\frac{dM}{dt} = K[(1-(M))]$$

$$\int\left(\frac{1}{(1-M)}\right).dM = K.\int_{t_0}^{t_1} dt$$

$$\ln(1-M) = K(t_1 - t_0)$$

$$\therefore M_i = \left(1 - e^{K(t_1-t_0)}\right) \dots\dots\dots\dots\dots\dots\dots\dots\dots\dots\dots\dots\dots\dots\dots\dots\dots3.6$$

This exponential relationship, using (M) as the mixing index, has been found to apply in many experimental investigations at least over two or three orders of magnitude of (M). In such cases, the constant K can be related to the mixing machine and to the conditions and it can be used to predict, for example, the times required to attain a given degree of mixing (Blackadder and Nedderman, 1971).

3.2.3 Energy consumption in mixing

Quite substantial quantities of energy can be consumed in some types of mixing, such as in the mixing of solids. There is no necessary connection between energy consumed and the progress of mixing. To use an extreme example there: could be shearing along one plane in a coarser material, then recombining to restore the original arrangement, then repeating which would consume energy but accomplish no mixing at all. However, in well-designed mixers energy input does relate to mixing progress, though the actual relationship has normally to be determined experimentally (Beek and Muttzall, 1975). The following equation is used to estimate the energy consumption requirement of the motor/stirrer for a specific mixing duty.

$$P = \frac{\left(\frac{I.V\sqrt{\phi.f.\eta}}{1000}\right).t}{m} \quad \text{...3.7}$$

P : Energy consumption(kW.h/kg)

I : Electrical current(Amperes)

V : Voltage(Volts)

ϕ : Phase

f : Power factor

η : Efficiency

t : time(hr)

m : mass of the blend (kg)

3.3 Experimental work

An experiment was conducted to evaluate the mixing characteristics of the 13.5 kg biomass blend consisting spent coffee grounds, coal fines, saw dust, chopped mielie husks and granulated paper, the sludge composition is shown in table 3.2. A

60 litre fixed speed mixer compete with an electrical starrier as shown in figure 3.5 and 3.6 was used to mix the material. While mixing, random samples were taken every 5 minutes for 40 minutes. These samples were measured for moisture contents using a direct heating moisture analyzer shown in figure 3.7. Values of the measured moistures per sample were recorded. The same procedure was repeated using coal fines of various particle sizes. Results of the blend containing the optimum coal fines particle size (80 %< 500µm) are presented in table 3.3. The mixing indices were calculated based on equation 3.12 and 3.13 as suggested by (Blackadder and Nedderman, 1971).The specific objectives of conducting these tests were to determine:

- The mixing constants
- Mixing indices
- Energy consumption during mixing.

Raw Material blend	Sample Mass(kg)	% (w/w)
Spent coffee ground	4.3	31.9
Finely ground coal fines 80%<250,500,1000 &2000µm	3.1	23.0
Saw dust	1.5	11.1
Mielie husk/grass clippings/cow dung	2.4	17.7
Shredded Paper(news print/card board)	1.4	10.3
Pulp containing water	0.8	5.9
Total material	13.5	100

Table 3.2: Proposed recipe for fuel briquette production

The same composition was used to conduct separate tests varying the particle size of coal fines in each blend.

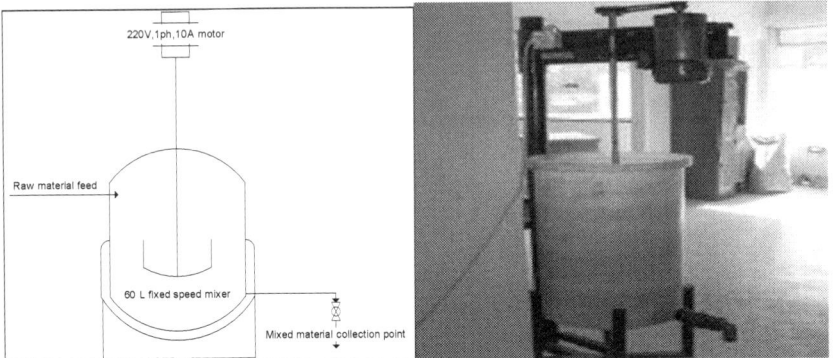

Figure 3.5 & 3.6: Fixed speed mixer complete with a stirrer used to conduct mixing tests.

Fixed speed mixer complete with a stirrer used to conduct mixing tests. The mixer is driven by a 220-volt, single phase electric motor with a power factor of 0.89, and the overall mechanical efficiency between motor and mixing blades is 75% as indicated in the by the manufacturers specifications.

Figure 3.7: Direct heating moisture analyser used for moisture content measurement in sample.

Samples were randomly drawn from the mixer and inserted into the moisture analyser, which directly heat up the containerised sample until all the water has been evaporated and gives displays the actual moisture content measured. These moistures were recorded and evaluated statistically for solid liquid uniformity which is a measure of mixing index.

3.3.1 Method

If particles are to be mixed, starting from various dry solids and water and ending up solids randomly distributed in sludge, the expected variances (S^2) of the sample composition from the mean sample (S) can be calculated. For a sludge mixture consisting of various solids with a fractional composition (k) and water fractional composition (w). If a reasonable number of representative samples are taken, it would be expected that the sum of fractional compositions of both solids and water would add up to 1.

$$k + w = 1 \dots\dots\dots\dots\dots\dots\dots\dots\dots\dots\dots\dots\dots\dots\dots\dots\dots\dots\dots\dots\dots\dots\dots\dots\dots\dots\dots\dots\dots\dots 3.8$$

- w: fractional composition of moisture in the sample
- k: fractional composition of solids in the sample

The mean water fractional compositions and variances are computed statistically using the following equations:

$$\overline{w} = \frac{1}{n}\left(\sum w\right) \dots\dots\dots\dots\dots\dots\dots\dots\dots\dots\dots\dots\dots\dots\dots\dots\dots\dots\dots\dots\dots\dots\dots\dots\dots\dots\dots 3.9$$

- n: number of samples
- $\overline{w}$: mean moisture fractional composition

$$S^2 = \frac{1}{n}\left(\sum_{i=0}^{n}\left((w_i)^2 - \left(\overline{w}\right)^2\right)\right) \dots\dots\dots\dots\dots\dots\dots\dots\dots\dots\dots\dots\dots\dots\dots\dots\dots\dots\dots 3.10$$

Using the probability theory the value of S_o^2 is computed as follows:

$$S_o^2 = \frac{1}{n}\left(pn(1-p)^2 + (1-p).n.(0-p)^2\right)$$

$$S_o^2 = p(1-p)\text{...}3.11$$

- p: probability of fractional moisture concentration in the sample to be a given value

It has been suggested by (Earle, 1986) that intermediate values between S_o^2 and S_r^2 could be used to show the progress of mixing. The suggestion is defined below assuming that S_r^2 is approximately zero as the number of the particles in the sample is very large. The mixing index (M_i) is a measure of mixing characteristics of material and its value and ranges between 0 and 1 during the course of mixing process.

$$M_i = \left(\frac{S_o^2 - S^2}{S_o^2 - S_r^2}\right)\text{...}3.12$$

The following equation is used to calculate the mixing constant (K) from the experimental data as suggested by (Beek and Muttzall, 1975) which is later used to determine the minimum required mixing time to attain material uniformity during mixing.

$$\frac{dM_i}{dt} = K[(1-(M_i))]$$

$$\int\left(\frac{1}{(1-M_i)}\right).dM_i = K.\int_{t_0=0}^{t_1=t} dt$$

$$\ln(1-M_i) = Kt$$

$$\therefore M_i = \left(1-e^{Kt}\right)\text{...}3.13$$

3.4 Results and discussions

Table 3.3 show the test results obtained after conducting mixing tests followed by drawing 10 random samples every 5 minutes interval. The methods described in section 3.3.1 were used to calculate other parameters such as power consumption, mixing constants and mixing indices.

Sample (#)	Mixing time (sec)	Moisture fraction (w)	$(w)^2 - (\overline{w})^2$	Power consumption $(kW.h/kg)$	Mixing Index (M_i)	$\ln(1 - M_i)$
1	0	0.650	-0.062	0.0000	0.000	
2	300	0.450	-0.282	0.0091	0.435	-0.570
3	600	0.770	0.108	0.0181	0.680	-1.140
4	900	0.540	-0.193	0.0272	0.819	-1.711
5	1200	0.860	0.255	0.0363	0.898	-2.281
6	1500	0.721	0.035	0.0453	0.942	-2.851
7	1800	0.740	0.063	0.0544	0.967	-3.421
8	2100	0.760	0.093	0.0635	0.982	-3.992
9	2400	0.712	0.023	0.0725	0.990	-4.562
10	2700	0.723	0.038	0.0816	0.994	-5.132
11	3000	0.730	0.048	0.0906	0.997	-5.702
Average	1500	0.696	0.012	0.045		

Moisture fraction probability in the sludge.	0.720
Average Mixing Index (M_i)	0.942
Average Mixing constant (k)	0.002
S^2	0.012
S_o^2	0.202
S_r^2	0.000

Legends:
Measured values
Calculated values

Table 3.3: Experimental data obtained during the mixing tests conducted with coal fines particles size 80 %< 500µm.

Detailed results for samples containing 1000, 2000 & 250 µm coal fines can be found in appendix D.

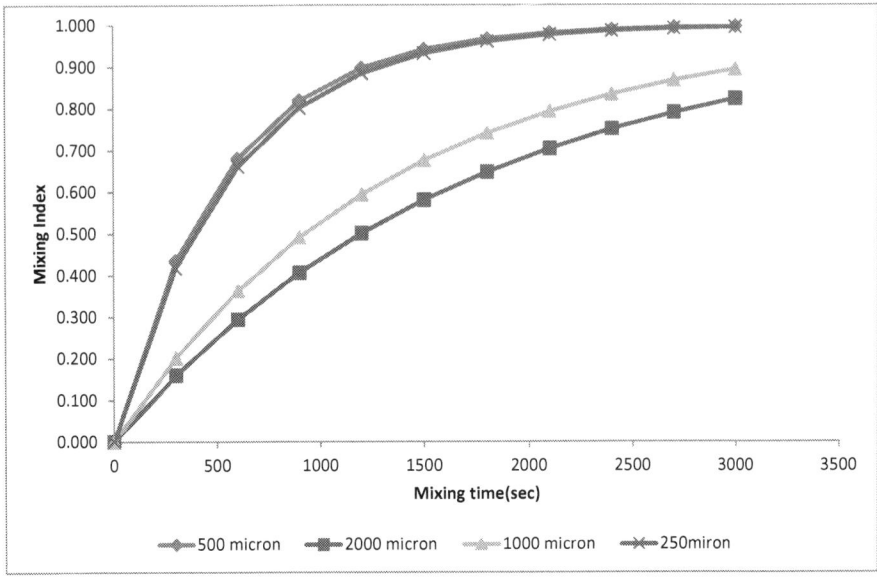

Figure 3.8: Variations of mixing index with mixing time for various coal fines particle sizes.

The aim of this test was to investigate the effect material size reduction to the mixing index of the sludge mix. Larger particles of coal fines in the mix tend to reduce stability of the briquettes when dried. This experiment investigates the relationship between the mixing indices and mixing times for samples prepared using the same compositions as indicated in table 1; however the coal fines were added at various particle size distributions i.e. 2000µm, 1000µm, 500µm and 250µm.

It was found that the when larger coal fines with an average particle size of 2mm were used in the mix, the mixing characteristics of the became poor as shown in figure 4 above and the mixing characteristics improved with finer coal fines particle being introduced in the mix. Better mixing is indicated by an increasing trend of mixing index with time; this behaviour is more visible in the material containing 500μm and 250μm coal fines in comparison to the material containing 2000μm and 1000μm coal fines particles. There is a significant effect of the coal fines particle size on the ease of mixing, this is clearly indicated differences in mixing indices between the material containing 2000μm and 1000μm coal fines particles.

Figure 5, shows manual screening of coal fines through 500 micron sieves. This preparation stage appears to be the most critical in terms of raw material preparation for the eco-fuel briquettes.

Figure 3.9: Manual sieving of coal fines through a 500μm screens at Thuthukani.

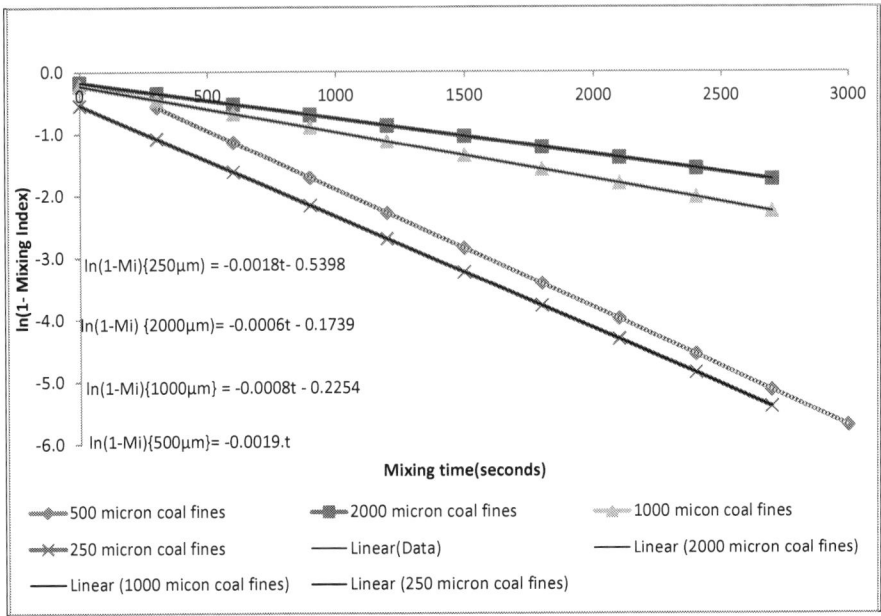

Figure 3.10: Graphical determination of the mixing constants of the specific material for the mixer used in the experiment.

With reference to equation 3.13, graphs of $\ln(1-M_i)$ versus mixing times were plotted on the same set of axis and linear trend lines were fitted in each graph. The graphs shows constant decreasing average slopes of:-

- $0.0019s^{-1}$ for blend containing 500µm coal fines.
- $0.018s^{-1}$ for blend containing 250µm coal fines.
- $0.008s^{-1}$ for blend containing 1000µm coal fines.
- $0.006s^{-1}$ for blend containing 2000µm coal fines.

This implies that the mixing constants increase when finer coal particles are added in the blend. It is also evident that mixing indices become better at higher mixing constants. When coal fines are ground below 250µm, it lowers the mixing index of the blend. This was noticed by a layer of froth formed during mixing. This effect is represented in figure 3.8; the blend containing 80% of coal fines below 500µm

38

reaches higher mixing indices compared to the blend containing 80% of coal fines below 250μm.

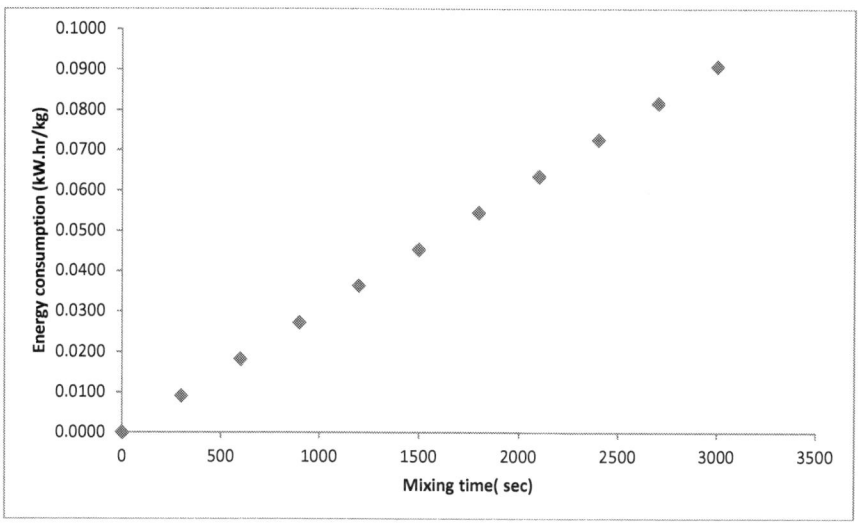

Figure 4: Graphical representation of the relationship between mixing time and energy consumption.

The energy consumption during mixing is calculated using equation 3.7, taking into account the mixer's technical specifications as stated out below figure 3.6. The results show a constant increase in energy consumption as time progresses. An average energy consumption of 0.04kW.hr/kg is noticed at an average mixing time of 25 minutes. This energy consumption is fairly low and could possibly be achieved by using a hand operated mixer. It is also clear that the shorter the mixing time, the less energy is consumed for that specific mixing duty. In principle the amount of energy required to achieve a certain mixing duty depends on the ability of the blend to reach a higher index over a short period of time.

3.4 Conclusions

The main conclusion of this experiment is that the coal fines particle size in the mix affects the mixing characteristics of the mixture as well as the energy requirements for mixing. Furthermore, if coarser coal fine particles are used, they create larger voids between the particles resulting in poor binding strength and reduced stability. The coal fines must be screened such that 80% of the fines used in the mixture pass through a 500μm sieve screen. The screening of the coal fine can be done manually as demonstrated in figure 5. When the correct particle size of coal fines are used, the mixture reaches uniformity quicker during mixing and less energy is consumed during mixing. The test results have shown that an average mixing energy of 0.04kW.hr per kilogram of material to be mixed is required. This is evidence that mixing can also be conducted manually, using a hand driven-mixer.

3.5 Reference

Alter, H. 1980. Recovery and re-use by mechanical processing. *Resource Recovery and Conservation*, Vol. 5, pp.73.

Barnes, N. 1985. Operational performance of waste-to-energy plants, *Seminar on Energy Recovery from Refuse Incineration:* Vol. 16(pp.28-32*)*, The Institution of Mechanical Engineers.

Beek, W. J. & Muttzall, K. M. K. 1975. *Transport Phenomena*, London: Wiley.

Blackadder, D. A. and Nedderman, R. M. 1971. *A Handbook of Unit Operations*, London: Academic.

Earle, R.L. 2006. *Unit Operations in Food Processing*, New York: Wiley.

Jackson, F. R. 1974. *Energy from Solid Waste*, England: Noyes Data Corporation.

Teeuwen, T. 1980. Introduction to the E. C. congress: packaging recovery and reuse, Resource *Recovery and Conservation*, Vol. *5,* pp. 3.

Wills B.A. & Napier-Munn, T.J. 2006. *Mineral Processing Technology*, Butterworth: Heinemann. pp.111.

3.6. Nomenclature and Units

Symbol	Definition	Units
W_i	Bond work index	$k.W.h/ton$
L	Particle size	μm
E	Energy	J
K_R	Rittinger's constant	–
x_1	Initial particle size	μm
x_2	Final particle size	μm
M_i	Mixing index	–
K	Mixing constant	sec^{-1}
t	Mixing time	sec
P	Unit energy consumption	$kW.h/kg$
I	Electrical current	A
V	Voltage	V
ϕ	Phase	–
f	Power factor	–
η	Efficiency	$\%$
W_{sieve}	Mass of sample retained on the sieve	$\%$
W_{Total}	Total mass of material screened	$\%$
m	Mass of blend	kg

CHAPTER 4: Eco-fuel briquettes production

4.1 Background

Briquetting of biomass is a densification process which improves its handling characteristics, enhances its volumetric calorific value, reduces transportation cost and produces a uniform, clean, stable fuel or an input for further refining processes. The briquetting of fuel was a significant business in the early 20th century. A multitude of plants around the world produced literally tens of millions of tons of usable and economic material that met the energy needs of industry. This process faded mid-century though, as other fuel sources became available (Granada, Lopez Gonzelez, Miguez & Moran, *2002*).

Fuel briquettes are bonded by the random alignment of fibers, generated when plant fibres and shredded waste paper and soaked in water. The process occurs at ambient temperature at a pressure of 1.5 to 3.0 MPa. To a large degree, the bonding force in the fuel briquette is mechanical, not chemical. Because of this, retaining fiber integrity and the right degree of plasticity in the mixture is crucial to the quality of the fuel briquette (Husain, Zainac & Abdullah, 2002).

Briquetting of biomass has been found to be a viable technology for upgrading biomass materials, including agricultural residues, particularly in developing countries where there are abundant biowaste resources. The technology converts the bio-waste into forms which are combustible in typical burners. The physical characteristics and, hence, combustion characteristics of the briquettes formed depend on several factors among which the briquetting pressure is controlled. This was confirmed by experimental investigations during which the samples were densified under pressure ranges of 5–15 MPa (Mani, Tabil & Sokhansanj, 2006).

4.2 Theoretical basis and method

4.2.1 Pressure density relationship

The relationship between the briquetting pressure and briquette density has been studied by many researchers in the past. O'Dogherty & Wheeler (1984) proposed a relationship between pressure and density for straw in one form of a simple power law at high briquetting pressures. This power law was developed for solid briquettes with diameters ranging between 40mm and 60mm. O'Dogherty & Wheeler have found a different exponential relationship of the form:

$$D = a \ln P + b \dots\dots\dots 4.1$$

Where P is the briquetting pressure measured in MPa, D is the density of the briquettes in kg/m^3 and a, b are empirical constants which vary for different feed stocks.

The constants for briquetting straw obtained by O'Dogherty & Wheeler are as follows:

40mm diameter, $a = 0.0389$, $b = 0.0045$,

50mm diameter, $a = 0.0871$, $b = 0.0036$,

60mm diameter, $a = 0.189$, $b = 0.0033$,

Faborode & O'Callagham (1986) expanded O'Dogherty & Wheeler's work and also noticed an exponential relationship between the briquetting pressure and density of the briquette when briquetting is conducted at moderately low pressure. This relationship is more relevant to the eco-fuel briquettes due to the fact that it was developed on circular shaped briquettes with a hole at the centre

$$P = ae^{bD} \quad \text{..4.2}$$

4.2.2 Axial load and pressure calculations in a screw press

The minimum force exerted by a screw mechanism is mainly dependent on the geometry of the screw and applied toque.

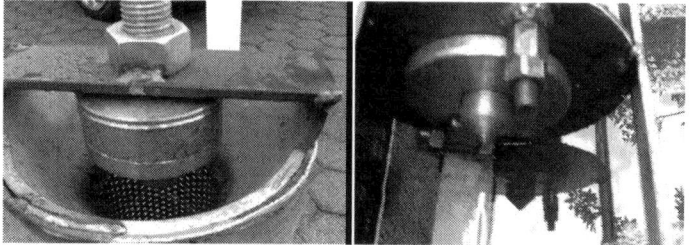

Figure 4.1: Plan view of the screw press equipment used for making eco-fuel briquettes.

The biomass blend is fed through the funnel of the press into the perforated cylinder, which is closed with a cam-lock solid steel disc at the bottom as shown in figure 4.1. A rotational force applied on the handle of the screw rod is transferred into an axial load, pushing the piston against the material in the cylinder. This force is maintained by the nut and as the screw rod is tuned, the axial load increases periodically.

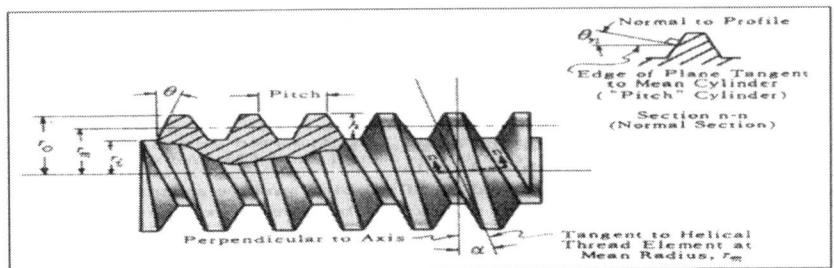

Figure 4.2: Cross section of a screw thread (source: Engineer's Edge World Wide Web, retrieved 26 January 2010. http://www.engineersedge.com/gears/screw-axial-thrust-load-calculations.htm)

The specifications of the screw thread used in the screw press are as follows:

- Applied toque (T) =27,935N.mm
- Major thread diameter (D) =30mm
- Thread radius (r_o) = 15mm
- Thread depth (h) = 3.8mm
- Coefficient of screw thread and mating (f) =0.15
- Thread pitch (L) = 3.74mm
- Thread angle at bearing surface (θ_n) =15^0

The following equations are used to calculate the axial load of the screw press based on the methods stated in Engineer's Edge World Wide Web, retrieved 26 January 2010. (http://www.engineersedge.com/gears/screw-axial-thrust-load-calculations.htm).

Minor thread radius:

$$r_i = r_o - h \dots\dots\dots\dots\dots\dots\dots\dots\dots\dots\dots\dots\dots\dots\dots\dots\dots\dots\dots\dots\dots\dots\dots\dots\dots\dots\dots\dots 4.3$$

Mean thread radius:

$$r_m = \left(\frac{r_o + r_i}{2}\right) \dots\dots\dots\dots\dots\dots\dots\dots\dots\dots\dots\dots\dots\dots\dots\dots\dots\dots\dots\dots\dots\dots 4.4$$

Angle of thread at mean radius:

$$\alpha = \tan^{-1}\left(\frac{L}{2.\pi.r_m}\right) \dots\dots\dots\dots\dots\dots\dots\dots\dots\dots\dots\dots\dots\dots\dots\dots\dots 4.5$$

Thread angle at bearing surface:

$$\theta_n = \tan\theta.\cos\alpha \dots\dots\dots\dots\dots\dots\dots\dots\dots\dots\dots\dots\dots\dots\dots\dots\dots\dots\dots\dots\dots\dots 4.6$$

Thread constant:

$$R_c = r_m \times \left[\left(\frac{\dfrac{\tan(\alpha)_{rad} + f}{\cos(\theta_n)_{rad}}}{\dfrac{1 - f.\tan(\alpha)_{rad}}{\cos(\theta_n)_{rad}}}\right) + f\right] \dots\dots\dots\dots\dots\dots\dots\dots\dots\dots\dots 4.7$$

Minimum axial load

$$\therefore F_{\min} = \frac{T}{R_c} \dots\dots\dots\dots\dots\dots\dots\dots\dots\dots\dots\dots\dots\dots\dots\dots\dots\dots\dots\dots\dots\dots 4.8$$

The briquetting pressure is calculated from first principles using the following equation.

$$P = \frac{F}{A_c}$$ 4.9

F = Force perpendicular to the cross section area of the briquette (N)

A_c = Cross sectional area of the briquette (m^2)

P = Briquetting pressure (Pa)

4.3 Experimental

A briquetting experiment was conducted using the screw press, Porta press and the Bikernmayer press. The main purpose of this experiment was to measure and calculate the maximum possible briquetting pressure in each press. This would allow other parameters such as cycle times, empirical constants, moisture contents and densities of the briquettes to be measured for each press. The briquettes were produced using the proposed recipe as stipulated in table 2.2 of chapter 2 of this dissertation. A measured quantity of the prepared sludge was poured into the screw press funnel and compacted until the final briquette is discharged as shown in figure 4.3. The maximum weight equivalent to the force exerted on to the handle of the screw press was measured and this force was recorded.

Figure 4.3: Pictorial representation showing how briquettes are made using a screw press.

The weight equivalent to the force exerted on to the Porta press was measured by pressing the briquettes on to a scale and this was also recorded.

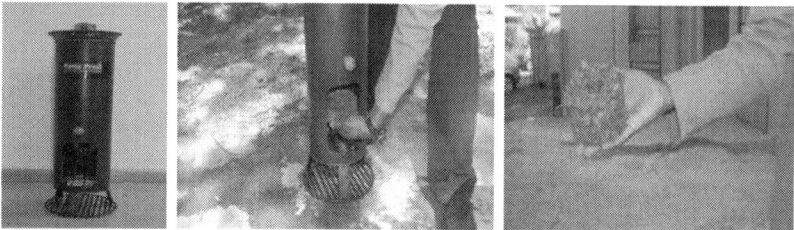

Figure 4.4: Pictorial representation showing how briquettes are made using a Porta press.

The tests were conducted at Bikernmayer facilities using a laboratory scale Bikernmayer press whereby parameters such as pressing pressure, moisture and briquettes densities were measured and recorded.

Figure 4.5: Pictorial representation showing how briquettes are made using a laboratory scale Bikernmayer press.

4.4 Results and discussions

4.4.1 The Screw press

The force applied on to the screw handle by one hand may be estimated as 22.6 kg weight equivalent, including the mass of the handle. This result in a weight of 221.7 N, which is equivalent to the tangential force exerted on the handle. Based on the general arrangement drawing in figure 4.6, a resultant force of 221.7 N is applied at perpendicular distance from the origin. The resulting toque may be calculated as follows:

$T = 221.7\,N \times 126\,mm = 27,935\,N.mm$

The axial load was calculated based on the procedure stipulated in section 4.4.2, using equations 4.3 to 4.9.

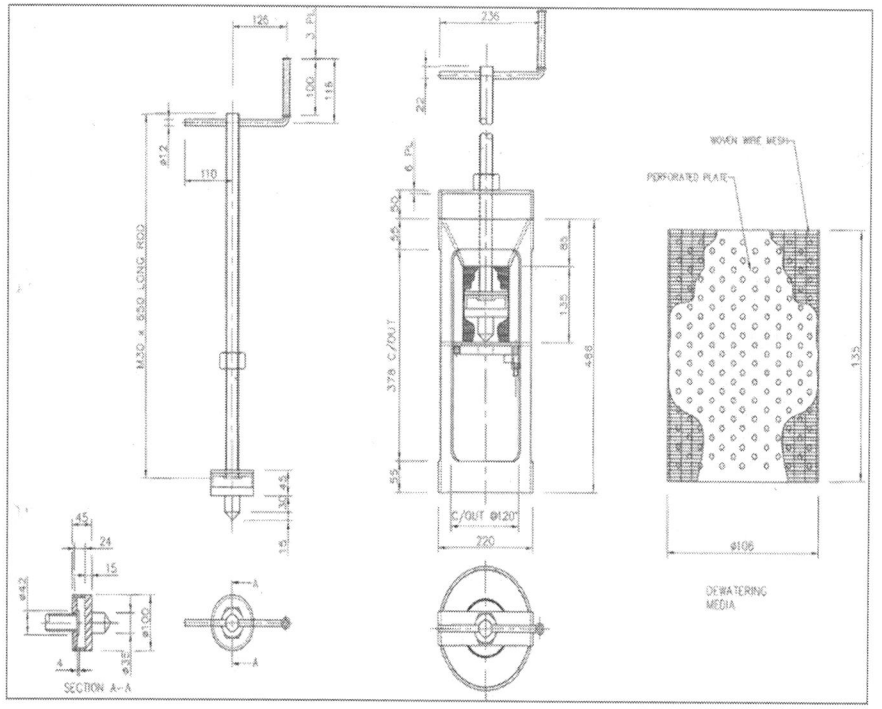

Figure 4.6: General arrangement drawing of the mechanical screw press (Drawn by J. Chivavaya using AutoCAD).

The results show the axial load parallel to the thread axis of 6,051 N when the screw friction coefficient is 0.15. If the screw thread is greased, the friction coefficient will be reduced and the maximum axial load can be computed as follows, neglecting friction effects.

$$W = F.d$$
$$\therefore W = F \times \pi d_i$$
$$\therefore W = 221.7 N \times 256 \times 10^{-3} m = 56.75 N.m$$

$$\therefore F_{max} = \frac{W}{L} = \underline{\underline{15,173.8 N}}$$

The pressure exerted on to the material is calculated as follows:

$$A_c = \frac{\pi(D_i^2 - d_i^2)}{4} = \frac{\pi(0.1m)^2 - (0.035m)^2}{4} = 6.89 \times 10^{-3} m^2$$

$$\therefore P_{min} = \frac{F_{min}}{A} = \frac{6,051N}{6.89 \times 10^{-3} m^2} = 878,229 Pa \approx \underline{\underline{0.878 MPa}}$$

$$\therefore P_{max} = \frac{F_{max}}{A} = \frac{15,173.8N}{6.89 \times 10^{-3} m^2} = 2,202,292.7 Pa \approx \underline{\underline{2,2 MPa}}$$

The pressure attained by the screw press falls within the range of 1.5-3.0 MPa as stipulated by Granada et al. (2002). However the quality of the briquettes was good due to the binding characteristics of the partially decomposed material. The briquettes produced by the screw press appeared to be more compact when compared to the ones produced by the Porta and Bikernmayer presses. This is justified by the quantity of the pressure exerted by the screw mechanism (0.878 MPa) as well as the dewatering characteristics provided by the compaction chamber of the press. Although the pressure achieved by the screw press is slightly lower that the pressure indicated by Granada et al. (2002), the briquettes were more compact and durable.

4.4.2 The Bikernmayer press

The Bikernmayer press was tested using a lab scale brick press that produces a single hexagonal briquette per cycle. The tests were conducted at Bikernmayer facilities whereby parameters such as pressing pressure, moisture and dry bulk densities were measured. The prototype has been used to conduct an economic evaluation of the batch production of eco-fuel briquettes based on the extrapolated results obtained from the lab scale brick press. Although the unit that produces four standard shape briquettes has not been constructed and tested, Bikernmayer has a proven technology on the prototype that produces 4-6 briquettes per cycle with various brick shapes.

Based on the test results obtained from the lab scale press, the expected pressure delivered by the proposed prototype is estimated as follows: The average pressure obtained in the lab scale press is 3.01MPa, this is to be transferred over four (4) standard briquettes with a larger surface area.

$$F_t = PA = 3.01 \times 10^6 \, N.m^{-2} \times 6.89 \times 10^{-3} \, m^2 = 20{,}738.9N$$

$$\therefore F = \frac{20{,}738.9N}{4} \, 5{,}184.7N$$

$$\therefore P = \frac{5{,}184.7N}{6.89 \times 10^{-3} m^2} = \underline{0.753MPa}$$

The expected moisture achievable in the proposed prototype Bikernmayer press is estimated by extrapolating the average measured moisture of 26.86% as obtained from the lab scale press.

$$\% \, moisture = \left(1 + \frac{0.753 \, MPa}{3.01 \, MPa}\right) \times 26.86\% \approx 33.58\%$$

4.4.3 Porta press

The maximum amount of pressure that can be exerted on a Porta press is calculated based on the average body mass that could be exerted on to the press. Given the mass of mass of 72 kg measured during briquetting, the average briquetting pressure is obtained as follows:

$$P = \frac{72 \ kg \ .9.81 \ m.s^{-2}}{6.89 \times 10^{-3} \ m^2} = 0.11 \ MPa$$

This pressure is far below the ones suggested in the literature as stipulated by (Granada et al., 2002).The total cycle time required to produce one briquette in a Porta press is 30 second, this is quicker than a screw press but the briquettes contained moistures of 44.9% and are the least compact when compared with the ones made from screw press and Bikernmayer press.

4.4.4 Overall discussions

Table 4.1 shows a summary of the briquetting pressure, density, moisture content and cycle times for all the equipments used in making eco-fuel briquettes.The results clearly indicate that the briquettes were made at moderately low pressure(below 5MPa) and therefore the Faborode & O'Callagham's (1986) power law is applicable to the pressure density relationship. A combination of the Screw press and Bikernmayer press measured variables were plugged into the power law in order to determine the empirical constants for the feed stocks used in making eco-fuel briquettes.

$$(3.01MPa) = a.e^{(860kg/m^3).b} \ \ldots\ldots\ldots\ldots\ldots\ldots\ldots\ldots for \ P = 3.01Mpa, \rho = 860kg.m^{-3}$$
$$(2.2MPa) = a.e^{(732kg/m^3).b} \ \ldots\ldots\ldots\ldots\ldots\ldots\ldots\ldots for \ P = 2.2Mpa, \rho = 732kg.m^{-3}$$

Solving the above equations simultaneously, the following empirical constants are obtained.

For a 100 mm diameter Eco- fuel briquettes,

$a = 0.363$

$b = 0.0025$.

These empirical constants look reasonable when compared to the ones determined by O'Dogherty & Wheeler (1984) in section 4.2.1.

	Screw press	Porta press	Bikernmayer lab scale	Prototype Bikernmayer
Pressure	0.878-2.2 MPa	0.11 MPa	3.01 MPa	0.753 MPa
Briquette density	732 kg/m^3	213 kg/m^3	860kg/m^3	645kg/m^3
Moisture content	28.9%	44.9%	26.8%	33.5%
Cycle time	3.5 min	0.5 min	0.17min	0.17min
Empirical constants for eco-fuel briquettes blend-Calculated using Faborode & O'Callagham 's model(1986)				
a	0.0025			
b	0.363			

Table 4.1: Summary of test results for various equipments used for making eco-fuel briquettes

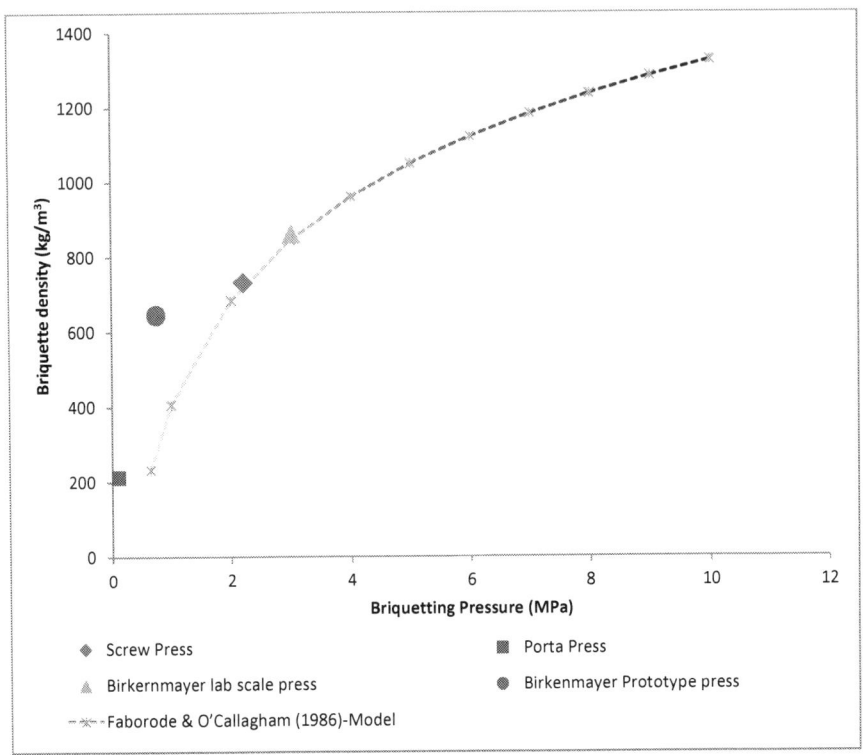

Figure 4.7: Graphical presentation of pressure & density relationship with reference to Faborode & O'Callagham's empirical model.

The results presented in figure 4.7 indicate that the briquettes made using the screw press and the Bikernmayer press follow the empirical model suggested by Faborode & O'Callagham (1986). This implies that if the material is pressed at pressures of 6- 8 MPa, briquettes of higher densities (1200kg/m^3) could be achieved.

The screw pressing mechanism has the largest pressing force over a given cross sectional area, resulting with final briquettes of higher densities (732 kg/m^3) and lower moisture content of 28.87%. The quality of briquettes produced by the screw press is superior compared to the ones produced by the Porta press. The main disadvantage of the screw press is that with its longer cycle time, fewer briquettes can be produced over a specified period of time and the operating principle of the press is labour intensive. On the other hand the Bikernmayer press is capable of producing briquettes of higher densities and lower moisture content compared to the screw press at slightly higher operating pressure. The main advantage of the Bikernmayer press its lowest cycle time compared to the other presses. The Prototype Bikernmayer press is the modified machine capable of processing four (4) briquettes per cycle, meaning that the production rate can be quadrupled and yet produce eco-fuel briquettes of reasonable densities and moisture content, in addition to that, the manufacturing cost of the equipment is lower.

Equipment capacity	Cycle time (min)	Briquettes mass (g)	Production rate(kg/h)	# briquettes/hr
4 x Screw press	3.5	200	13.71	69
4 x Porta press	0.5	200	96	480
1 x Bikernmayer press	0.17	200	282.35	1412

Table 4.2: Proposed production capacities for different equipment.

Table 4.2 shows the total cycle times obtained experimentally using the screw press, Porta press and the Bikernmayer press. The Bikernmayer has the shortest cycle time of 10.2 seconds compared to the Porta press of 30 seconds and screw press with the longest total cycle time of 210 seconds. The cycle time includes material pressing, dewatering and briquette discharge. Although the Bikernmayer press was tested on hexagonal shaped briquettes (each weighing 150g) as shown in figure 4.5, it is assumed that the average cycle time for standard shape briquettes will be approximately equal to the one obtained on the hexagonal bricks and the pressures might be less than the one obtained experimentally due to the larger surface area of the standard shape briquettes. The proposed Bikernmayer prototype press can produce up to 4 briquettes per cycle compared to the other presses with only produces 1 briquette per cycle. This makes the screw and Porta presses labour intensive as three extra units and operators are required to match the capacity of the Bikernmayer press.

4.5. Conclusions

The batch production of the eco-fuel briquettes was conducted using the Porta press, screw press and the Bikernmayer lab scale press. The samples were made from a sludge mixture containing, 31.91 % spent coffee beans, 23.04% finely grinded coal fines, 11.08% saw dust, 17.73% mielie husks, and 10.34% granulated paper and 5.91% pulp containing waste water. The above recipe may be varied depending on the availability of raw material. In principle, any biomass raw material can be blended accordingly to ensure sufficient binding. Based on the findings, the briquettes produced using the above mentioned equipment were of good quality, however the screw press had the longest cycle time of 3.5 minutes compared to the to the Porta press and Bikernmayer press with cycle times of 30 seconds and 10.2 seconds respectively. The proposed Bikernmayer model is capable of producing four eco-fuel briquettes per cycle which makes it four times greater than the Porta and screw presses in capacity. The fabrication cost of the Bikernmayer press is lower as compared to the other two presses. The operating pressures of the Porta, Bikernmayer and Screw presses are 0.11, 0.753 and 0.878 MPa respectively, and this is lower than the briquetting pressure stated in the literature 1.5- 3 MPa(Husain, Zainac & Abdullah, 2002). However, these pressures were obtained using electrical power, whereas a focus point in this project is to limit the dependency on electrical power. It has been observed that the briquettes made from the screw press were more stable and compact compared to the others with a density of 732 kg/m^3 and moisture content of 28.87%. It may be possible to accept a slightly less stable product in the light of the energy saving that was achieved.

The proposed prototype Bikernmayer press is thought to be best suited for batch production of eco-fuel briquettes due to its low cost, capacity and availability. The press has less moving parts which makes it safer to be operated by two people. The test results have proven that the briquettes can be sufficiently compacted without applying a significant amount of pressure. This was confirmed by visual inspection of the briquettes produced. The Bikernmayer press can operate at a pressure of 0.753 MPa and the briquettes produced have a density of 645kg/m^3 and moisture content of 33.48%.

4.6. References

Debdoubi A., El amarti A. & Colacio E., 2005. Production of fuel briquettes from esparto partially pyrolyzed. *Energy Conversion and Management*, 46: 1877-1884.

Demirbas A. & Sahin A., 1997. Evaluation of biomass residue: Briquetting waste paper and wheat straw mixtures. *Fuel Processing Technology*, 55: 175-183.

Engineer's Edge World Wide Web, retrieved 26 January 2010. http://www.engineersedge.com/gears/screw-axial-thrust-load-calculations.htm

Faborode M.O. & O'Callagham J.R., 1986. Theoretical analysis of compression of fibrous agricultural materials. *Journal of Agricultural Engineering*, 35: 175-9.

Granada E., Lopez Gonzelez L.M., Miguez J.L. & Moran J., 2002. Fuel lignocellulosic briquettes die design and product study. *Renewable Energy*, 27: 561-573.

Husain, Z., Zainac Z. & Abdullah Z. 2002. Briquetting of palm fibre and shell from the processing of palm nuts to palm oil. *Biomass and Energy*, 22: 505-509.

Mani, S., Tabil L.G. & Sokhansanj S. 2006. Specific energy requirement for compacting corn stover. *Bioresource Technology*, 97: 1420-1426.

Mouton J. 2001: *How to succeed in your Master's and Doctoral studies*: A South African Guide and Resource Book. Pretoria: Van Schaik.

O'Dogherty H.J. & Wheeler J.A., 1984. Compression of straw to high densities in close cylindrical dies. *Journal of Agricultural Engineering*, 29: 61-71.

4.7. Nomenclature and Units

Symbol	Definition	Units
A	Briquette cross sectional area	m^2
d	Circumferential distance	m
d_i	Briquette inner diameter	m
D_i	Briquette outer diameter	m
f	Friction coefficient of screw thread and mating	$-$
F	Force	N
F_{min}	Minimum axial force	N
F_{max}	Maximum axial force	N
h	Thread depth	mm
L	Thread pitch	mm
R_c	Thread constant	mm
r_i	Minor thread radius	mm
r_o	Major thread radius	mm
r_m	Mean thread radius	mm
T	Toque	$N.mm$
W	Work done	$J/N.m$
θ_n	Thread angle at bearing surface	$^0/rad$
α	Latent heat of water	$^0/rad$

CHAPTER 5: Drying

5.1 Background

Drying generally refers to the removal of a liquid from a solid by evaporation. Mechanical methods for separating a liquid from a solid are not generally considered drying, although they often precede a drying operation, since it is less expensive and frequently easier to use mechanical methods than to use thermal methods (Treybal, 1987). Drying involves the final removal of relatively small amounts of water from material. For example, a moisture content of 10-20% by volume would normally allow particles to flow freely, yet suppress dust formation. The necessity for drying may be to make a product suitable for sale (e.g. paint pigments), or for subsequent processing for example in pyrometallurgical operations (McCabe and Smith, 1956).

Drying is usually the final step in the working cycle and generally, it follows the filtration or centrifugation step and precedes the grinding step. When a solid dries, two fundamental and simultaneous processes occur: (1) heat is transferred to evaporate liquid; (2) mass is transferred as a liquid or vapour within the solid and as a vapour from the surface. These factors governing the rates of these processes determine the drying rate (Treybal, 1987). Commercial drying operations may utilize heat transfer by convection, conduction, radiation, or a combination of these. Industrial dryers differ fundamentally by the methods of heat transfer employed. However, irrespective of the mode of heat transfer, heat must flow to the outer surface and then into the interior of the solid. Careful consideration of many factors is necessary in the final selection of the most suitable type of dryer for a particular application. Some of these factors are:

- Properties of the material being handled
- Drying characteristics of the material
- Flow of the material to and from the dryer
- Product qualities

5.1.1 Rate of Drying – Convection Drying

It is essential to know the rates of drying of solids achieved under different conditions. It may be necessary to avoid the maximum rate of drying if, for example, it results in shrinkage, surface hardening, surface crazing or other undesirable effects in the drying solids. When a solid is dried experimentally, data are usually obtained relating moisture content to time. Consider the drying of a non-porous, insoluble material such as biomass fuel briquette. The surface of the briquette is exposed to a drying medium such as hot dry air passing over the surface. Figure 5.1 shows a typical drying curve (Perry & Green, 1998).

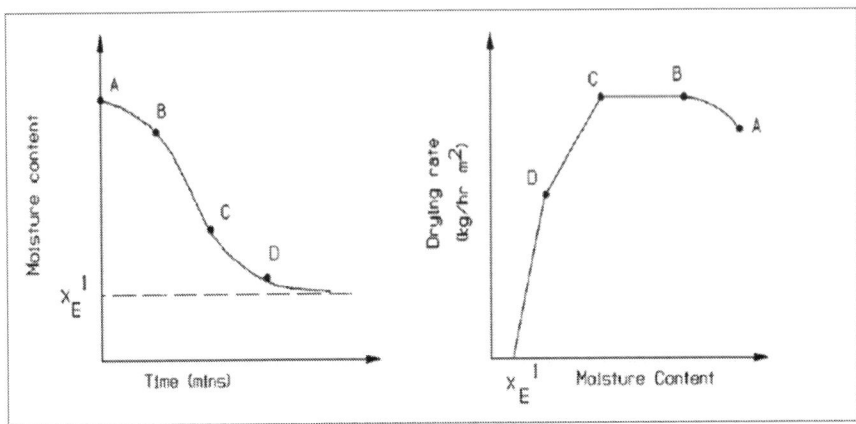

Figure 5.1: Typical Drying Curves (Source: *Perry & Green, 1998)*

Immediately after contact between the wet solid and the drying medium, the solid temperature adjusts until it reaches a steady state. The solid temperature and the rate of drying may increase or decrease to reach the steady state condition. The section *AB* on each curve represents a warming-up period of the solids. Section *BC* on each curve represents the constant-rate period (Perry, Green and Molony, 1997)

Point *C,* where the constant rate ends and the drying rate begin falling, is termed the critical-moisture content. The curved portion *CD* on Fig. 5.1 is termed the falling-rate period, and is characterised by a continuously changing rate throughout the remainder of the drying cycle.

Point *E* on Fig. 5.1 represents the point at which all the exposed surface becomes completely unsaturated and marks the start of that portion of the drying cycle during which the rate of internal moisture movement controls the drying rate. Portion *CE* in Figure 5.1 is usually defined as the first falling-rate drying period; portion *DE,* as the second falling-rate period. In the constant-rate period moisture movement within the solid is rapid enough to maintain a saturated condition at the surface, and the rate of drying is controlled by the rate of heat transferred to the evaporating surface (*Perry & Green, 1998*). In most drying operations, water is the liquid evaporated and air is the normally employed purge gas. For drying purposes, a Psychrometric chart found very useful is that reproduced in Figure 5.2 below.

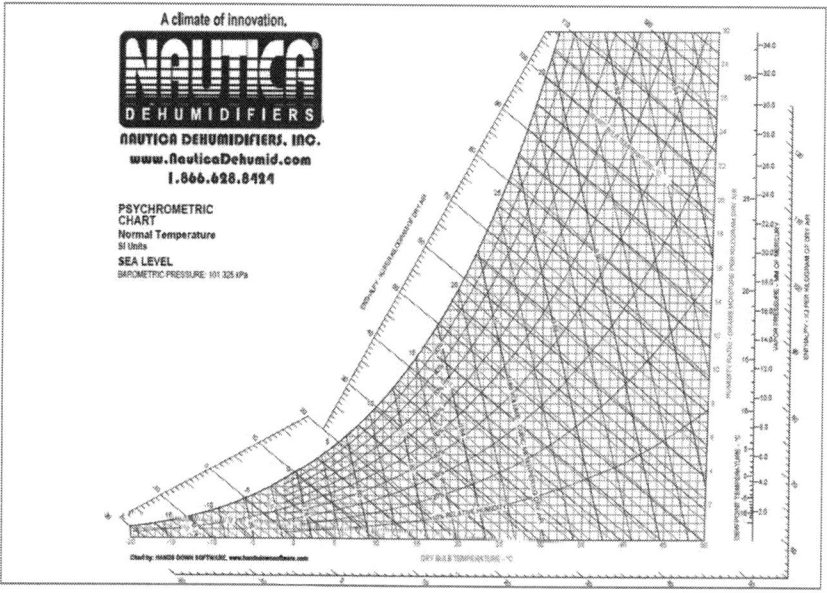

Figure 5.2: Psychrometric Chart at 101.325kPa (Source: *Perry & Green, 1998).*

1. The wet-bulb or saturation temperature line gives the maximum weight of water vapour that a kg of dry air can carry at the intersecting dry-bulb temperature.

2. The percent of relative humidity is defined by ratio of partial pressure of water vapour in the air, and vapour pressure of water at the same temperature.

3. Specific volumes are given by the curves entitled "Volume m³/kg dry air." The volumes are plotted as functions of absolute humidity and temperature. The

65

difference between dry-air specific volume and humid-air volume at a given temperature is the volume of water vapour.

4. Specific enthalpy data are given on the basis of kilojoules per kilogram of dry air and give an indication of the overall heat contained by the moist air under the given conditions.

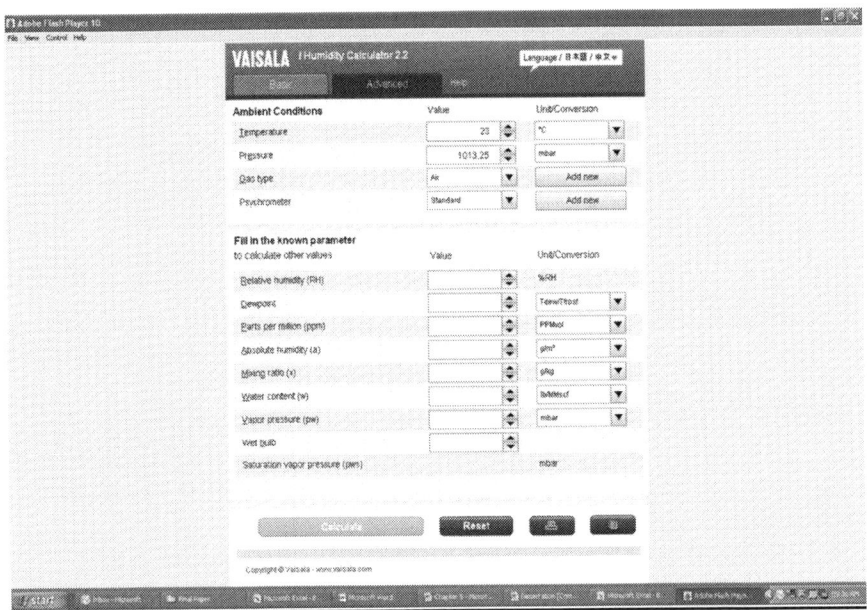

Figure 5.3: The Vaisala Humidity calculator 2.2 screen capture.

The above software is capable of calculating the relative humidity, dew point temperature, gas concentration and so forth, at a given ambient temperature and pressure using the pre-defined built in functions. It provides a flexibility of conducting experiments under non-standard conditions and still able to estimate other variable accurately. Figure 5.4 shows the results obtained when entering the gas temperature of 184.6^0C and pressure of 82.96 kPa.

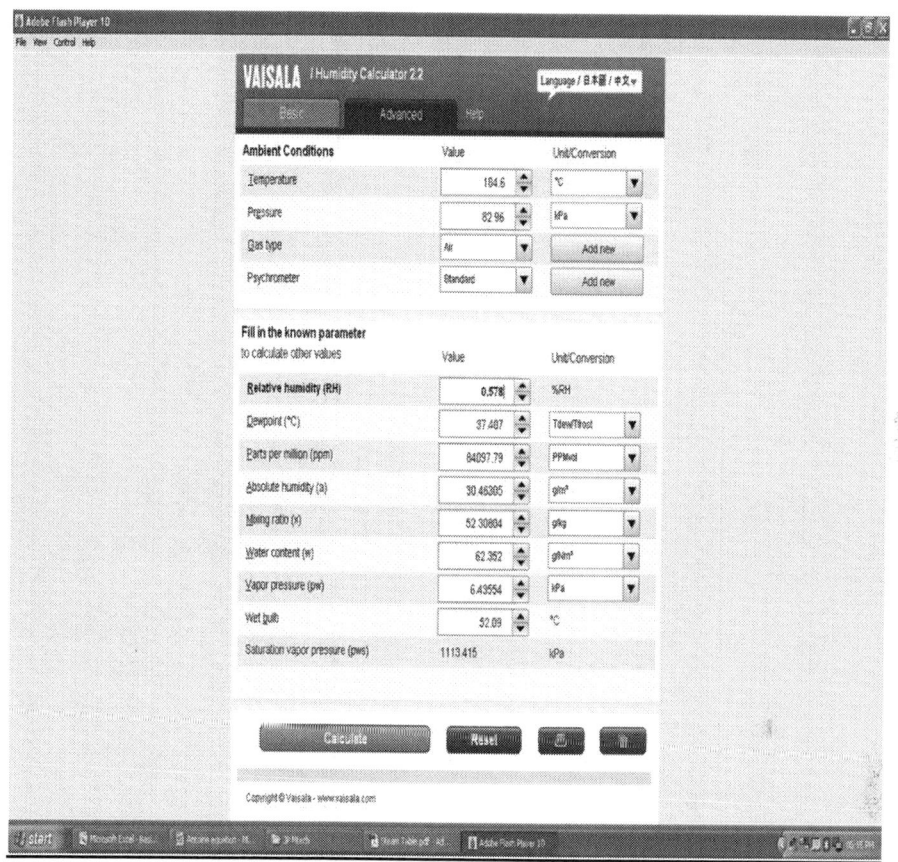

Figure 5.4: The Vaisala Humidity calculator 2.2 screen capture with results

5.3 Theory and Method

5.3.1 Liquid diffusion and mass transfer

Diffusion-controlled mass transfer is assumed when the vapour or liquid flow conforms to Fick's second law of diffusion. This is stated in the unsteady-state-diffusion equation using mass-transfer notations as shown in equation below (Treybal, 1987).

$$N_A = -D_{AB} \frac{dC_A}{dl} \dotfill 5.1$$

- C_A = Concentration of component A in a two-component phase of A and B,
- N_A = diffusion rate
- l = Distance in the direction of diffusion
- D_{AB} = Binary diffusivity of the phase AB.

This equation applies to diffusion in solids, stationary liquids, and stagnant gases. The diffusion resistances for various geometries are shown in figure 5.5 below. Equation 5.1 is further manipulated, allowing for geometry to yield the following equation:

$$N_A = \frac{C_{A1} - C_{A0}}{\dfrac{\ln\left(\dfrac{D}{d}\right)}{2.\pi l.D_{AB}}} \dotfill 5.2$$

The reference values for the diffusion coefficient of water vapour in air are best represented by Bolz and Tuve (1976) empirical model (*Perry & Green, 1998*).

$$D_{AB_T} = -2.77 \times 10^{-6} m^2.s^{-1} + \left(4.479 \times 10^{-8} m^2.s^{-1}.K^{-1}\right)T + \left(1.656 \times 10^{-10} m^2.s^{-1}.K^{-2}\right)T^2 \dotfill 5.3$$

Diffusion equations may also be used to study vapour diffusion in porous materials. It should be clear that all estimates based on relationships that assume constant diffusivity are approximations. Liquid diffusivity in solids usually decreases with moisture concentration. Liquid and vapour diffusivity also change, and material shrinks during drying.

Geometry	Species Concentration Distribution, $x_A(x)$ or $x_A(r)$	Species Diffusion Resistance, $R_{m,dif}$
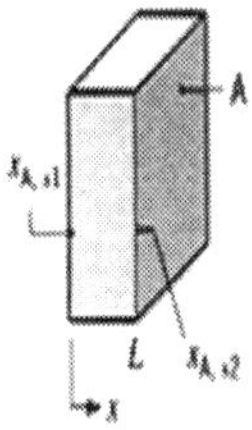	$x_A(x) = (x_{A,s2} - x_{A,s1})\dfrac{x}{L} + x_{A,s1}$	$R_{m,dif} = \dfrac{L}{D_{AB}A}$ [b]
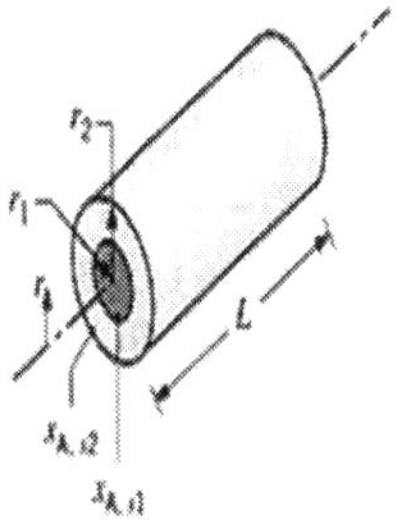	$x_A(r) = \dfrac{x_{A,s1} - x_{A,s2}}{\ln(r_1/r_2)}\ln\left(\dfrac{r}{r_2}\right) + x_{A,s2}$	$R_{m,dif} = \dfrac{\ln(r_2/r_1)}{2\pi L D_{AB}}$ [c]
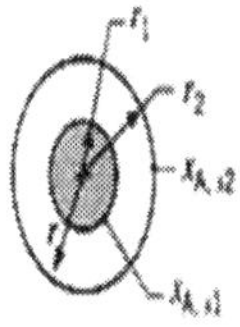	$x_A(r) = \dfrac{x_{A,s1} - x_{A,s2}}{1/r_1 - 1/r_2}\left(\dfrac{1}{r} - \dfrac{1}{r_2}\right) + x_{A,s2}$	$R_{m,dif} = \dfrac{1}{4\pi D_{AB}}\left(\dfrac{1}{r_1} - \dfrac{1}{r_2}\right)$ [c]

[a]Assuming C and D_{AB} are constant.
[b]$N_{A,x} = (C_{A,s1} - C_{A,s2})/R_{m,dif}$
[c]$N_{A,r} = (C_{A,s1} - C_{A,s2})/R_{m,dif}$

Figure 5.5: Summary of species diffusion solutions for stationery media Source (*Incropera & DeWitt, 1934*).

5.4 Experimental

The experiment was conducted using the setup shown in figure 5.6 below. The aim of this test was to determine the drying profile and energy requirements of the optimum eco-fuel briquettes under controlled conditions. The experiment was conducted under a constant pressure, temperature and relative humidity of 0.7845 atmospheres, 22.4 ^{0}C and 26.7% respectively. The unit consists of a tunnel in one end of which is mounted an axial fan. Downstream of the fan, a system of electrical resistances heats the air flowing to the drying chamber. The incoming ambient air is heated and dried to a lower relative humidity. The air velocity is measured using a digital air anemometer and the velocity can be varied by controlling the fan speed. The initial relative humidity is measured just after the air has been preheated and the final air humidity is measured after drying the briquette sample.

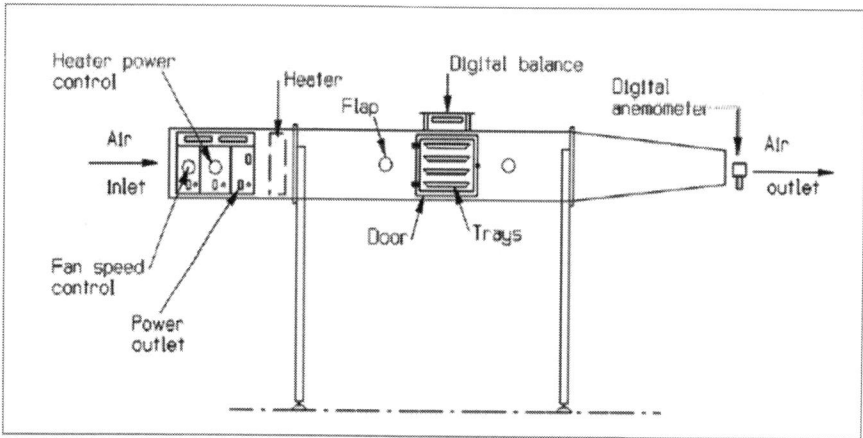

Figure 5.6: Automatic convective dryer (Martinus, 2007)

A representative sample of partially decomposed biomass sludge containing spent coffee grounds, coal fines, mielie husks and saw was collected in order to measure the amount of dry solids in the sludge (composition of this sludge is shown in figure 3.2). The sample was split into four crucibles, each weighing approximately 5 grams. Each crucible was put into an electronic moisture analyser for 15 minutes. The dry solids contents of the crucibles were compared and the average solids/ moisture contents were recorded. Part of the sludge was dried in an oven at 45^{0}C for two days and the total moisture /solids content was compared with the one obtained from the digital moisture analyser.

A pyknometer was used to measure the dry solids density of the sample, of which the density was found to be 1.081g/ml as compared to the calculated value of 1.075g/ml. 200ml of the biomass sludge weighing 204.13g was poured into a screw press, and pressed for 3.5 minutes. A total filtrate of 95.6 ml was collected during pressing and was found to weigh 95.64g. The briquette was discharged from the press and weighed before it can be put into a moisture seal bag.

The dryer was switched on and allowed to run for 15 minutes without heating. A hygrometer was used to calibrate the measuring instruments on the dryer. The ambient conditions were measured and the heater was switched on. The fan speed and heating power were adjusted and the system was allowed to run for 30 minutes for the air conditions to stabilise.

- A wet briquette was removed from the moisture seal bag and fixed at centre of the drying region, perpendicular to the air flow.
- The initial wet mass was recorded and the timer was started.
- Relative humidity, temperatures and wet mass were recorded every ten (10) minutes for seven (7) hours.

- The results were captured onto an excel spreadsheet for further analysis and calculations.

5.5 Results and discussions

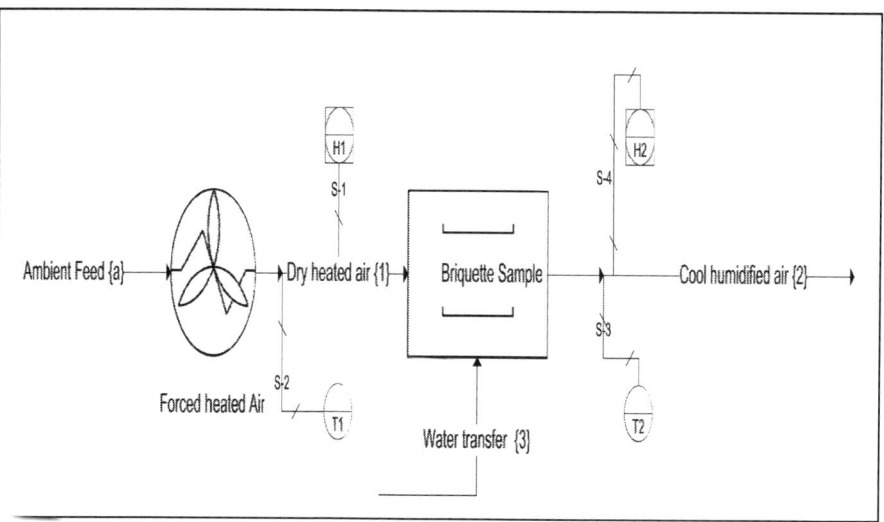

Table 5.7: Process Flow Diagram of the test setup shown in figure 5.6(Drawn using Microsoft Vision Ver. 2003).

The following basic calculations were performed using average values to indicate how mass and energy balance has been carried out. The following results were obtained after drying four biomass sludge samples (5g each) independently in a moisture analyser shown in figure 3.7., and the following moisture contents were measured in each sample.

Sample Number	Moisture content (w/w %)
1	71.80%
2	72.01%
3	72.31%
4	72.26%
Average	72.10%

Table 5.1: Moisture test results from a moisture analyser.

Confirmation tests were conducted by drying four sludge samples (30g each) in an oven at 45^0C for two days and the following results were obtained.

Sample Number	Moisture content (w/w %)
1	70.30%
2	72.40%
3	71.30%
4	73.00%
Average	71.75%

Table 5.1: Moisture test results from oven.

$$\therefore Dry_{solids} = 100\% - \left(\frac{72.10w/w\% + 71.75w/w\%}{2}\right) = 28.078w/w\%$$

The dry solids were measured for density using a pyknometer and a density of 1.081g/ml was obtained.

5.5.1 Mass and energy balance

The values used to conduct energy balance calculations are the averages over the whole experiment, detailed actual experimental data can be found in D. All the symbols used in this section are fully defined in section 5.8

5.5.1.1 Energy required evaporate water from the briquette sample

The amount of heat required to evaporate the moisture from the briquette sample is calculated adding the latent heat and sensible heat as follows:

$$Q = \overset{\bullet}{m}_{moisture} \left[(C_P . \Delta T) + \lambda \right]$$
$$Q = 1.81 \times 10^{-6} \, kg \, .s^{-1} \left[(4.18 \, kJ . kg . K^{-1} (8.89 \, K)) + 2438 \, .kJ \, / \, kg \right]$$
$$Q = 4.48 \, W$$

5.5.1.2 Energy absorbed by hot air

$$Q = \overset{\bullet}{m}_{d.a} \left[H_{in} - H_{out} \right]$$
$$= 0.0020 \, kg \, .s^{-1} \left[(50.03 \, kJ \, / \, kg) - (43.21 \, kJ \, / \, kg) \right]$$
$$= 13.64 \, W$$

The total heat transfer coefficient to allow for the 21.39W of energy to be lost e.g. from the wall of the tunnel to the air surrounding the apparatus is about 10.73 W/m^2.K as calculated below, which is the expected range for free convection.

$$Q_{loss} = \frac{T_{\infty,tunnel} - T_{\infty,ambient}}{\dfrac{1}{hA}}$$

$$\therefore h = \frac{Q_{loss}}{A.\Delta T} = \frac{21.39W}{4 \times 0.45m \times 0.45m \times 0.26m (30.76^0 C - 21.3^0 C)} = \underline{\underline{10.73W \, / \, m^2 .K}}$$

5.5.1.3 Energy transferred to heat up the air

$$Q = \overset{\bullet}{m}_{d.a} \left[H_{in} - H_a \right]$$
$$= 0.0020 kg.s^{-1} \left[50.037 kJ \, / \, kg - 37.1 kJ \, / \, kg \right]$$
$$= \underline{\underline{25.87W}}$$

5.5.1.4 Volumetric flow rate of active moist air

Given the cross sectional area (A) of the tunnel and the air velocity (v), the air volumetric velocity is calculated as follows:

$$\dot{V} = A.v$$

$$\dot{V}\ 0.00689m^2 \times 0.320m.s^{-1}$$

$$\therefore \dot{V} = \underline{0.0022048m^3.s^{-1}}$$

5.5.1.5 Mass flow rate of dry air

The mass flow rate of dry air remains constant throughout the streams, and is calculated below based on the conditions of stream (2) in figure 5.7 above.

$$\dot{V} = \left[\left(\frac{\dot{m}_{d.a}}{\rho_{d.a-out}}\right) + \left(\frac{\dot{m}_w}{\rho_w}\right)\right] = \left[\frac{\dot{m}_{d.a}}{\left(\frac{PM_a}{RT_{out}}\right)} + \frac{H_{out}.\dot{m}_{d.a}}{\rho_w}\right]$$

$$\therefore \dot{m}_{d.a} = \left[\frac{\left(\frac{\dot{V}.\rho_w.P.M_a}{RT_{out}}\right)}{\left(\rho_w + \left(\frac{P.M_a.H_{out}}{R.T_{out}}\right)\right)}\right] = \left[\frac{\left(\frac{(0.0022048m^3.s^{-1})\times(997.70kg.m^{-3})\times(0.7845.atm)\times(0.0289kg.mol^{-1})}{(8.21\times10^{-5}.m^3.atm.K^{-1}.mol^{-1})\times(298.9K)}\right)}{(997.70.kg.m^{-3}) + \left(\frac{(0.7845.atm).(0.0289kg.mol^{-1}).(0.0022048)}{(8.21\times10^{-5}\,m^3.atm.K^{-1}.mol^{-1})\times(298.9K)}\right)}\right]$$

$$\therefore \dot{m}_{d.a} = 0.002037kg.s^{-1}$$

5.5.1.6 Water balance

The mass flow rate of moisture in the heated air plus the moisture absorbed by the air from the wet briquette is equal to the moisture contained by the cooled air leaving the dryer.

$$\dot{m}_1 + \dot{m}_3 = \dot{m}_2$$

$$\therefore \dot{m}_{d.a}(H_1) + \dot{m}_{d.a}(H_3) = \dot{m}_{d.a}(H_2)$$

$$\therefore H_3 = H_2 - H_1$$

$$\therefore H_3 = \left(\frac{6.58g_w}{kg_{d.a}}\right) + \left(\frac{5.71g_w}{kg_{d.a}}\right) = 0.00087\,kg_w.kg_{d.a}^{-1}$$

$$\dot{m}_3 = 0.002037\,kg_{d.a}.s^{-1}.(0.00087\,kg_w.kg_{d.a}^{-1}) = 0.00177219\,g_w.s^{-1}$$

$$Total\_mass\_loss = 0.00177219\,g_w.s^{-1} \times (420 \times 60)s = 44.65g$$

The experiment shows the total mass loss of 43.57g which indicates an experimental error of 2.44%.

5.5.2 Drying curves

Wet briquettes were dried in a convective oven; data relating moisture content to time were obtained. The data plotted as moisture content (wet basis) x *(w/w %)* versus drying time t (minutes), as shown in Figure 5.8. This curve represents the actual drying case when a wet solid briquette loses moisture from the initial moisture content of 46.78 w/w% first by evaporation of the free moisture from a saturated surface on the wet briquette until the state of equilibrium is reached. Figure 5.8 indicates that the drying rate is subject to variation with time and free moisture content. The warm-up drying period occurs in the first 100 minutes of the drying cycles, followed by a constant rate period for the next 220 minutes. The system characteristics indicate the equilibrium moisture content of 10.73w/w%.

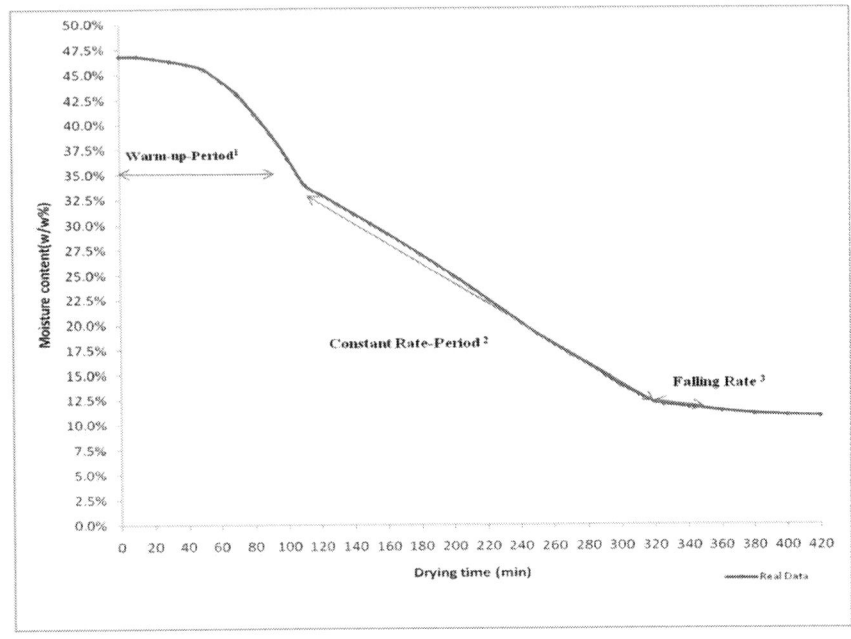

Figure 5.8: Drying curve; moisture content vs. drying time.

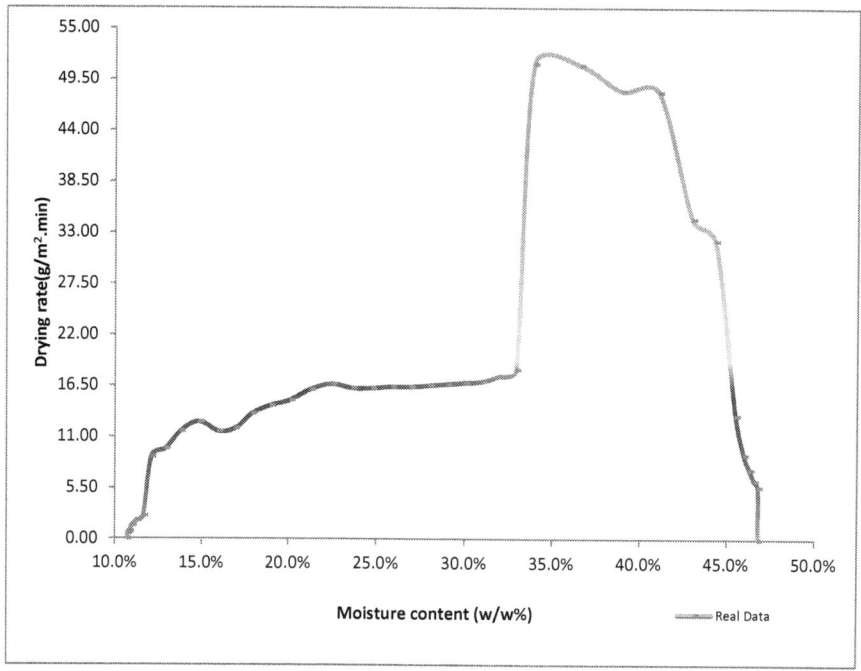

Figure 5.9: Drying curve: Drying rate vs. moisture content.

The above graph indicates a steep rise in drying rate during the warm-up period with a peak of 51.23g/m^2.min at 33.93 % moisture before the constant drying rate of 16.84g/m^2.min is attained. The average drying rates for the drying phases were recorded as follows:

- Warm-up : $\partial_1 = 25.64 g.m^{-2}.min^{-1}$
- Constant rate: $\partial_2 = 14.77 g.m^{-2}.min^{-1}$
- Falling rate: $\partial_3 = 1.67 g.m^{-2}.min^{-1}$

The drying rate drops to zero at equilibrium with a possibility of further drying beyond bound moisture, should the drying conditions be altered. Based on the test results, the mass percent of moisture removed at each stage is estimated as follows:

- Warm up period:- $\dfrac{46.78\% - 33.92\%}{36.05\%} = 35.67\%$

- Constant rate period:- $\dfrac{33.92\% - 12.11\%}{36.05\%} = 60.50\%$

- Falling rate period:- $\dfrac{12.11\% - 10.73\%}{36.05\%} = 3.83\%$

The required total drying time was calculated by simply adding the times for each of the three drying phases as follows:

$$t(\text{min}) = \frac{m_b(g).x(w/w\%)}{A(m^2)}.\left[\frac{0.3567}{\partial_1(g.m^{-2}.\min^{-1})} + \frac{0.605}{\partial_2(g.m^{-2}.\min^{-1})} + \frac{0.0383}{\partial_3(g.m^{-2}.\min^{-1})}\right]$$

$$t = \frac{107.32g \times 0.4715}{0.00689m^2}.\left[\frac{0.3567}{25.64g.m^{-2}.\min^{-1}} + \frac{0.607}{14.77g.m^{-2}.\min^{-1}} + \frac{0.0383}{1.67g.m^{-2}.\min^{-1}}\right]$$

$$t = 7,344.18g.m^{-2} \times \left(0.0774g^{-1}.m^2.\min\right)$$

$$\therefore t = 572\,\text{min}\,(\text{convection only})$$

5.5.3 Effect of mass transfer to drying

Figure 5.10 shows a mass transfer boundary layer through an eco-fuel briquettes while it is been dried in a convective drier. The direction of air flow is from A0 to A1 as shown in the diagram below. The moisture concentration in each phase was calculated by taking the reciprocals of the specific volumes of the air in each phase. The mass transfer of water from the briquettes to the air is driven by the concentration gradient between phase A0 and A1 and the mass transfer resistance as defined by Perry and Green (1998) in equation 5.1 and 5.2.

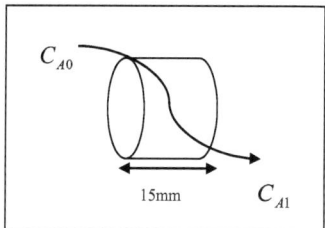

Figure 5.10: Mass transfer boundary layer through a dried briquette

$$C_{A0} = \frac{1}{v} = \frac{1}{1.10 m^3/kg} = 0.904 kg/m^3$$

$$C_{A1} = C_{sat@T_2}$$

The mass transfer rate depends on the diffusivity of moisture in air which is defined by Bolz and Turve (1976) as a function of temperature in equation 5.3.Based on Bolz and Turve (1976) modified empirical model, the air-water diffusivity is estimated as follows.

$$D_{AB_T} = -2.77\times10^{-6} m^2.s^{-1} + (4.479\times10^{-8} m^2.s^{-1}.K^{-1})T + (1.656\times10^{-10} m^2.s^{-1}.K^{-2})T^2$$

$$T = \left(\frac{410.25K + 295\ 55K}{2}\right) = 352.90K$$

$$D_{AB_T} = -2.77\times10^{-6} m^2.s^{-1} + (4.479\times10^{-8} m^2.s^{-1}.K^{-1})352.9K + (1.656\times10^{-10} m^2.s^{-1}.K^{-2})\times(352.90K)^2$$

$$\therefore D_{AB_T} = 3.486\times10^{-6} m^2.s^{-1}$$

Characteristics length and convective heat transfer area contributes to the resistance of mass transfer (Treybal, 1987) and is calculated based on the dimensions of the briquette as follows:

$$L = \left(h + \left(\frac{D-d}{2}\right)\right) = \left(15\times10^{-3}\ m + \left(\frac{100\times10^{-3}\ m - 35\times10^{-3}\ m}{2}\right)\right) = 0.0475\ m$$

$$A = \left[\frac{2\pi \left(D^2 - d^2 \right)}{4} + \pi \left(D + d \right)h \right]$$

$$= \left[\frac{2\pi \left((0.1m)^2 - (0.035\,m)^2 \right)}{4} + \pi (0.1m + 0.035\,m) \times 15 \times 10^{-3} \right] = 1.58 \times 10^{-2}\,m^2$$

The properties of air-water vapour were evaluated in order to derive a generic mass transfer rate equation that depends on the saturation concentration in phase A1. The saturation concentration depends on saturation temperature which is calculated by an iterative process.

Properties of air-water vapour:

$$T = 352.9\,K\,; \nu = 2.446 \times 10^{-6}\,m^2.s^{-1}; D_{AB} = 3.486 \times 10^{-6}\,m^2.s^{-1}; Sc = \frac{\nu}{D_{AB}} = \left(\frac{2.446 \times 10^{-6}\,m^2.s^{-1}}{3.486 \times 10^{-6}\,m^2.s^{-1}} \right) = 0.703$$

Reynolds Number

$$Re_L = \frac{VL}{\nu} = \frac{0.21\,m.s^{-1} \times 0.0475\,m}{15.89 \times 10^{-5}\,m^2.s^{-1}} = 62.7$$

Sherwood Number

$$Sh_L = 0.664 \sqrt{Re_L} \sqrt[3]{Sc}$$

$$Sh_L = 0.664 \times (62.7)^{\frac{1}{2}} \times (0.703)^{\frac{1}{3}}$$

$$Sh_L = 4.67$$

$$Sh_L = \frac{h_m.L}{D_{AB}}$$

$$\therefore h_m = \frac{4.67 \times 3.486 \times 10^{-6}\,m^2.s^{-1}}{0.0475\,m} = 3.427 \times 10^{-4}\,m.s^{-1}$$

Mass transfer rate for drying eco-fuel briquette defined as follows

$$N_A = h_m.A.(C_{A0} - C_{A1})$$

$$N_A = 3.427 \times 10^{-6}\,m.s^{-1} \times 1.58 \times 10^{-2}\,m^2 \times (0.940\,kg.m^{-3} - C_{sat})\dots\dots\dots\dots\dots\dots\dots\dots\dots\dots\dots\ 5.4$$

An experiment was conducted by burning eco-fuel briquettes in a ceramic stove through a chimney whereby the following parameters were recorded in order to calculate the saturation temperature and concentration.

$$T_{flame} = 137.1\,^{\circ}C$$
$$T_s = 22.4\,^{\circ}C$$
$$\overline{T}_f = 79.75\,^{\circ}C \; ; \; Air$$
$$\rho = 0.9805 \; kg.m^{-3}$$
$$\mu = 210.76 \times 10^{-7} \; kg.m^{-1}.s^{-1}$$
$$v = 21.75 \times 10^{-6} \; m^2.s^{-1}$$
$$k_f = 30.44 \times 10^{-3} \; W.m^{-1}.^{\circ}C^{-1}$$
$$\alpha = 35.6 \times 10^{-6} \; m^2.s^{-1}$$
$$\beta = 2.73 \times 10^{-3} \; K^{-1}$$
$$Pr = 0.69883$$
$$L = 47.5 \; mm$$

Energy balance calculations for convection and radiation were carried out as follows, using the relevant empirical correlations as proposed by Incropera and De Witt (1934).

$$Ra_L = Gr_L . Pr = \frac{g.\beta(T_s - T_\infty).L^3}{v.\alpha}$$

$$\therefore Ra_L = \frac{9.81 m.s^{-2} \times 2.73 \times 10^{-3} K^{-1}.(410.25K - 295.55K) \times (47.5 \times 10^{-3} m)^3}{21.75 \times 10^{-6} m^2.s^{-1} \times 35.6 \times 10^{-6} m^2.s^{-1}}$$

$$Ra_L = \frac{3.29 \times 10^{-4} m^4.s^{-2}}{7.74 \times 10^{-10} m^4.s^{-2}}$$

$$\therefore Ra_L = 4.25 \times 10^5$$

$$\overline{Nu}_L = \left\{ 0.825 + \frac{0.387 \, Ra_L^{1/6}}{\left[1 + (0.492 / Pr)^{9/16}\right]^{8/27}} \right\}$$

$$\overline{Nu}_L = \left\{ 0.825 + \frac{0.387 \left(4.25 \times 10^5\right)^{1/6}}{\left[1 + \left(\frac{0.492}{0.69883}\right)^{9/16}\right]^{8/27}} \right\}$$

$$\therefore \overline{Nu}_L = 3.64$$

$$h_{out} = \frac{\overline{Nu}_L \times k_f}{L} = \left(\frac{3.64 \times 30.44 \times 10^{-3} W.m^{-1}.K^{-1}}{47.5 \times 10^{-3} m}\right) = 2.33 W.m^{-2}.K^{-1}$$

83

$$\overline{Nu} = \frac{\overline{h}L}{k_f} = (0.037 \, \text{Re}_L^{4/5} - 871) \, \text{Pr}^{1/3}$$

$$\therefore \text{Re}_L = \left(\frac{\left(\dfrac{2.33}{\sqrt[3]{0.69883}} + 871 \right)}{0.037} \right)^{\frac{5}{4}}$$

$$\therefore \text{Re}_L = 2.93 \times 10^5 \prec 5 \times 10^5 \, (Invalid)$$

$$\overline{Nu} = \frac{\overline{h}L}{k_f} = 0.0296 \, \text{Re}_L^{4/5} \, \text{Pr}^{1/3}$$

$$\therefore \text{Re}_L = \left(\frac{\overline{Nu}}{0.0296 \sqrt[3]{\text{Pr}}} \right)^{\frac{5}{4}} = \left(\frac{3.64}{0.0296 \sqrt[3]{0.69885}} \right)^{\frac{5}{4}} = 475 \, (Valid)$$

The outside plate velocity was calculated as follows:

$$u_\infty = \frac{\text{Re}_L \, \mu}{\rho.L} = \frac{475 \times 210.76 \times 10^{-7} \, kg.m^{-1}.s^{-1}}{0.9805 \, kg.m^{-3} \times 47.5 \times 10^{-3} \, m} = 0.21 \, m.s^{-1}$$

Heat transfer by convection and radiation on the Outside, assuming black body radiation.

$$q = q_{conv} + q_{rad}$$
$$\dot{m}\lambda = q_{conv} + q_{rad}$$
$$h_m.A_1.(C_{A0} - C_{sat@T_S})\lambda = hA(T_{flame} - T_s) + \varepsilon\delta A(T_{flame}^4 - T_s^4)$$
$$\therefore C_{sat@T_S} = \left[C_{A0} - \frac{hA(T_{flame} - T_s) + \varepsilon\delta A(T_{flame}^4 - T_s^4)}{h_m.A_1.\lambda} \right]$$
$$\therefore C_{sat@T_S} = \left[0.914 kgm^{-3} - \frac{2.33Wm^{-2}.K^{-1}.1.58\times10^{-2}m^2(41025K - T_s) + 1.5.67\times10^{-8}Wm^{-2}.K^{-4}.6.89\times10^{-3}m^2((41025K)^4 - T_s^4}{3.427\times10^{-4}ms^{-1}.1.58\times10^{-2}m^2.2426\times10^3 \, J.kg^{-1}} \right.$$

The above equation is solved numerically be choosing any value of T_s and calculate the corresponding saturation concentration, the resulting saturation concentration is compared with the value on the steam table. It was found that the saturation concentration of 0.03396 kg/m³ matched the saturation temperature of 31.85 °C on the steam table. Based on the above obtained target iteration values, the heat transfer calculations are carried out as follows.

$$q = q_{convQ} + q_{rad}$$
$$\dot{m}.\lambda = q_{conv} + q_{rad}$$
$$q = hA_1\left(T_{flame} - T_s\right) + \varepsilon\delta A\left(T_{flame}^4 - T_s^4\right)$$
$$q_{conv} = 2.33\,W.m^{-2}.K^{-1}.1.58\times 10^{-2}\,m^2.(410.25\,K - 305\,K)$$
$$\therefore q_{conv} = 3.87\,W \approx (33.5\%)$$
$$q_{rad} = 5.67\times 10^{-8}\,W.m^2.K^4.6.89\times 10^{-3}\,m^2.\left((410.25\,K)^4 - (305\,K)^4\right)$$
$$\therefore q_{rad} = 7.69\,W \approx (66.5\%)$$
$$\therefore q = 11.56\,W$$

$$\phi = \frac{11.56\,W}{0.01\,m^2} = 1{,}156\,W/m^2$$

If the drying heat transfer fluxes of 1,156 W/m² is used for each briquette with an exposed surface area of 0.01m². The heat transfer efficiency of 87.75% is used based on the combustion test results.

$$n_b = \frac{1m^2}{0.01m^2} = 100$$

$$E = 1{,}156.09\,J.s^{-1} \times 289\,min \times 60\,s \times 0.8775 = 17.58\,MJ$$

$$\therefore m_b(req) = \frac{17.85\,MJ}{18.9\,MJ.kg^{-1}} = \left(\frac{0.944\,kg}{118\times 10^{-3}\,kg}\right) \approx 8\,brique\,ttes$$

This confirms that 8 dry briquettes each weighing 118g can be burnt to provide sufficient heat energy to dry 100 wet briquettes each weighing 200g. The evaporation rate resulting with the heat energy calculated above (11.56W).

$$E_{rate} = \frac{q}{\lambda} = \frac{11.56 \times 10^{-3} \, kJ.s^{-1}}{2426 \, kJ.kg^{-1}} = \underline{\underline{4.765 \times 10^{-6} \, kg.s^{-1}}} \approx \left(0.285 \, g.\min^{-1}\right)$$

In simple terms, 0.944 kg of dry briquettes can be burnt to produce sufficient heat required to dry 20 kg of wet briquettes. The following data was used in the economic model.

- Total production rate: 49,694 kg/month
- Total number of briquettes: 248,470 briquettes/month(each wet briquette weighing 200g)
- Daily production rate: 11,295 briquettes/day (normal shifts of 22 days per month)
- Hourly production rate: 1,412 briquettes/hour (normal shift of 8 hour per day)

The total number of briquette dried in a daily shift: 11,295 briquettes per cycle

Surface drying area: $A_t = 1,412 \times 0.01 m^2 \approx \underline{\underline{65 m^2}}$

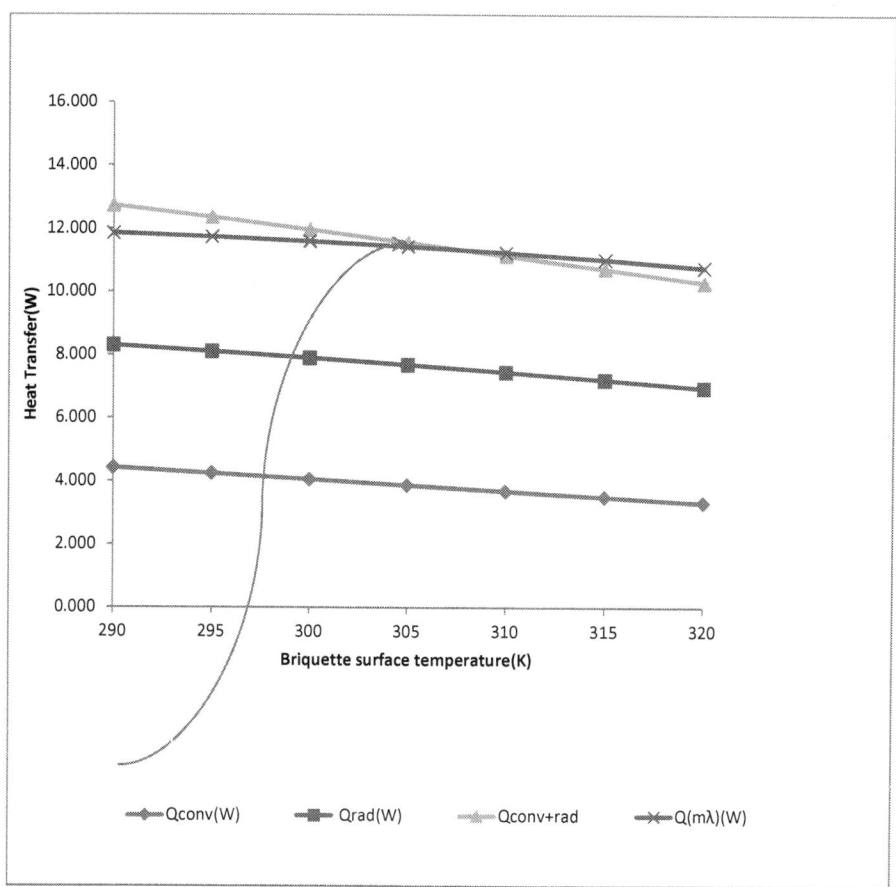

Figure 5.11: Variations of drying heat transfer with briquette surface temperatures and the saturation temperature.

5.5.4 Effect of direct heat drying on drying time

The required drying time for eco-fuel briquettes was estimated using the initial moisture content after pressing and drying rates as obtained from the laboratory drying profiles as indicated in figure 5.9. Considering that the briquettes are dried to an equilibrium moisture content of 10.73%.

Given the mass of wet briquette as 200g after pressing, with an initial moisture content of 33.48%, the required drying time was computed as follows:

$$m_b(wet) = 200g; m_b(dry) = 118g$$
$$x = 33.48\%;$$

$$t(min) = \frac{m_b(g).x(w/w\%)}{A(m^2)}.\left[\frac{0.3567}{\partial_1(g.m^{-2}.min^{-1})} + \frac{0.605}{\partial_2(g.m^{-2}.min^{-1})} + \frac{0.0383}{\partial_3(g.m^{-2}.min^{-1})}\right]$$

$$t = \frac{200\,g \times 0.3348}{0.0158\,m^2}\left[\frac{0.3567}{25.64\,g.m^{-2}.min^{-1}} + \frac{0.607}{14.77\,g.m^{-2}.min^{-1}} + \frac{0.0383}{1.67\,g.m^{-2}.min^{-1}}\right]$$

$$t = 4,237.97\,g.m^{-2} \times \left(0.0774\,g^{-1}.m^2.min\right)$$

$$\therefore t = 328\,min(\text{Covection only})$$

$$\therefore t = \left(\frac{(200-118)\times 10^{-3}\,kg}{4.765 \times 10^{-6}.kg.s^{-1}}\right) = 289.99\,min(\text{radiation \& convection}) \approx 11.5\%\text{ reduction in time}$$

The above calculations show that it would take 5.5 hours to dry a 200g briquette from 33.48% initial moisture to 10.73% equilibrium moisture via convection. This can be compared with the 9.5 hours taken in the laboratory to dry 107.32g with higher initial moisture content of 47.1% as shown under section 5.8. This indicates that the initial moisture content is an important factor in the drying time and cost. However, if briquettes are burnt to provide heat energy for drying, this will reduce the overall drying time by 11.5% since more heat is supplied by radiation resulting in the total drying time of 4.8 hours. It was found from the briquettes combustion tests that each briquette will burn for approximately 3.5 hours, which makes the drying process more practical by recovering the waste heat energy to pre dry the briquettes. The following calculations shows the amount of heat that can be transferred when eco-fuel briquettes are combusted, the same heat will eventually be used to speed up the drying process.

5.6 Conclusions

The drying characteristics of the fuel briquette in an automatic convective dryer were investigated. The drying pattern follows the one described by Perry & Green, 1998. Although the drying test was conducted for 9.5 hours before all the drying regimes may be identified, it was noticed that 35.67% moisture is driven off during the warm-up period, followed by 60.5% moisture reduction during the constant rate period and 3.83% during falling rate period. The warm-up and constant rate period takes 5.3 hour at this stage the briquette is dried to about 12.5% moisture. In the refractory brick fireplace, a drying heat requirement flux of 1,156 W/m^2 is expected, which includes the radiation component (66.5% of total). This implies that 1,156W will be required to dry 100 briquettes to 10.73% moisture within a period of 4.8 hours. The most economical method is to use the fuel briquette as a source of energy for the dying process, as it costs less than R1.06/kg to produce eco-fuel briquettes with an energy value of 11.9c per MJ for a briquette, while the electricity price is 12.5c per MJ (before the 2010 increases). The average briquette surface temperature during drying is estimated to be 32 ^0C, which is reasonable for convection and radiation drying at atmospheric pressure and this will also prevent driving off the volatiles from the eco-fuel briquettes and possible ignition. Based on the combustion finding discussed in chapter 6 of this dissertation, eight dry briquettes each weighing 118grams at a net calorific value of 18.9 MJ/kg and heat transfer efficiency of 87.6%, will provide 17.58 MJ of heat energy. If this is energy is to be transferred over 4.6 hours drying time, it will yields 1,156W drying heat rate. In simple terms, eight eco-fuel briquettes may be used to dry 100 briquettes.

The heat transferred by the briquettes can evaporate water at a rate of 0.285g/min. The proposed drying facility is similar to a fire place with refractory bricks on the inside and base floor to prevent heat loss to the surroundings. The wet briquettes will be dried on the steel grid placed perpendicular to the burning flames of the briquettes like a braai. Other wet briquettes can be stacked in a multi tray above the drying briquettes to improve energy efficiency by using the waste heat to pre dry the wet briquettes. The required drying area suitable for the capacity of 1,412 briquettes per day is $65m^2$ .The test results obtained in this work forms a basis of economic evaluation discussed later in chapter 7 of this dissertation.

5.7 References

Coulson, J. M. and Richardson, J. F., 1978.*Chemical Engineering, 3rd Edition*, Vols.2, Pergamon, Oxford.

Fishenden, M.D. and Saunders, O. A. 1950. *Introduction to Heat Transfer*, Oxford University Press, Oxford.

Incropera, F.P & DeWitt, D.P R., 1934. *Fundamentals of Heat and Mass transfer*, 5th Edition, John Wiley and sons .ISBN: 0-471-28650-2

McCabe, W. L., Smith, J. C. and Harriott, P. 1975. *Unit Operations of Chemical Engineering*, 3rd Edition, McGraw-Hill, New York.

Perry R & Green D, 1998. *Perry's Chemical Engineer's handbook*. New York: McGraw-Hill, pp 986, 1998.

Perry, R. H., Green, D.W. and Molony, J.D. 1997. *Chemical Engineers' Handbook*, 7th Edition, McGraw-Hill,

Treybal, R. E. 1987. *Mass Transfer*, 3rd Edition, McGraw-Hill, New York

5.8 Nomenclature and Units

Symbol	Definition	SI units
A_a	Active drying area	m^2
C_p	Specific heat capacity	$kJ.kg^{-1}.K^{-1}$
D_b	Outer diameter of the briquette	m
d_b	Inner diameter of the briquette	m
t_b	Thickness of the briquette	m
H_a	Ambient air relative humidity.	$w/w\%$
H_1	Heated air relative humidity.	$w/w\%$
H_2	Humidified air relative humidity.	$w/w\%$
$\bar{H}_a$	Ambient air absolute humidity.	$kg_{\cdot moisture}/kg_{\cdot dry.air}$
$\bar{H}_1$	Heated air absolute humidity.	$kg_{\cdot moisture}/kg_{\cdot dry.air}$
$\bar{H}_2$	Humidified air absolute humidity.	$kg_{\cdot moisture}/kg_{\cdot dry.air}$
T_a	Ambient air temperature.	^0C
T_1	Incoming hot air temperature.	^0C
T_2	Exiting humid air temperature	^0C
T_a	Ambient air temperature	^0C
ΔT	Temperature difference	^0C
T_m	Mid-point temperature	^0C
v_{ave}	Average air velocity	$m.s^{-1}$
$\dot{m}_{w.a(2)}$	Mass flow rate of humidified air	$kg.s^{-1}$
$\dot{m}_{w.(2)}$	Mass flow rate of moisture after drying	$kg.s^{-1}$
$\dot{m}_{d.a.(2)}$	Mass flow rate of dry air	$kg.s^{-1}$
m_b	Mass of the wet briquette	kg
M_a	Molecular mass of air	$kg.mol^{-1}$

$\hat{h}_a$	Specific enthalpy of ambient air	$kJ.kg^{-1}$
$\hat{h}_1$	Specific enthalpy of incoming heated air	$kJ.kg^{-1}$
$\hat{h}_2$	Specific enthalpy of exiting humidified air	$kJ.kg^{-1}$
h_{air}	Convective heat transfer coefficient of air	$W.m^{-2}.K^{-1}$

(Continued)

k_a	Thermal conductivity of air	$W.m^{-1}.K^{-1}$
$\hat{v}_a$	Specific volume of moist air	$m^{-3}.kg^{-1}$
P	Atmospheric pressure	Pa
R	Ideal gas constant	$m^3.Pa.mol^{-1}.K^{-1}$
Q_r	Thermal duty	$J.s^{-1}$
Q_l	Heat loss	$J.s^{-1}$
Q_s	Heat supplied	$J.s^{-1}$
x	Briquette moisture content	$w/w\%$

∂	Drying rate.	$g.m^{-2}.min^{-1}$
η	Dryer efficiency.	$\%$
λ_w	Latent heat of water	$kJ.kg^{-1}$
$\rho_{d.a}$	Density of dry air	$kg.m^{-3}$
$\rho_{m.a}$	Density of moist air	$kg.m^{-3}$
ρ_w	Density of water	$kg.m^{-3}$
μ_a	Viscosity of air	$kg.m^{-1}.s^{-1}$
Pr	Prandlt Number	Dimensionless
Nu	Nusselt Number	Dimensionless
Re	Reynolds Number	Dimensionless
ϕ	Heat flux	$W.m^{-2}$
E	Energy	J
E_{rate}	Evaporation rate	$g.min^{-1}$

CHAPTER 6: Combustion and Gas emissions

6.1 Theory-Background

Flue gas emissions from biomass combustion refer to the gas product resulting from burning of biomass solid fuel. Solid fuels are mostly burnt with ambient air as opposed to combustion with pure oxygen. Since ambient air contains about 79 volume percent gaseous nitrogen (N_2), which is essentially non-combustible, the largest part of the flue gas from most fossil fuel combustion is inert nitrogen. The next largest part of the flue gas is carbon dioxide (CO_2) which can be as much as 10 to 15 volume percent or more of the flue gas. This is closely followed in volume by water vapour (H_2O) created by the combustion of the hydrogen in the fuel with atmospheric oxygen. Much of the smoke seen pouring from flue gas stacks is this water vapour forming a cloud as it contacts cool air (Beek and Muttzall, 1975).

A typical flue gas from the combustion of fossil fuels will also contain some very small amounts of nitrogen oxides (NO_x), Hydrogen sulphide (H_2S), sulphur dioxide (SO_2) and particulate matter .The nitrogen oxides are derived from the nitrogen in the ambient air as well as from any nitrogen-containing compounds in the fossil fuel. The sulphur dioxide is derived from any sulphur-containing compounds in the fuels. The particulate matter is composed of very small particles of solid materials and very small liquid droplets which give flue gases their smoky appearance (Blackadder and Nedderman, 1971).

6.1.1 Combustion

Combustion is primarily defined as an exothermic process in which a fuel reacts with oxygen to give off energy in the form of heat and light.

Figure 6.1: The flame (heat and light) caused as a results of an eco- fuel briquette undergoing combustion (Picture taken at UJ FADA centre).

Combustion usually takes place between and fuel and air in the presence of some activation energy. In the large majority of real world uses combustion, air is the main source of oxygen. Air contains approximately 79 and 21 mol % of Nitrogen and Oxygen respectively. This means for every kilogram of Oxygen in air there is 3.76 kg of Nitrogen. It is expensive to obtain Oxygen in its pure form, thus most combustion processes are conducted in excess air. The resultant flue gas from the combustion will normally contain Nitrogen and some emissions due to the nature of hydrocarbon containing impurities (Henderson and Perry, 1976).

6.1.2 Measurement of calorific value

A Bomb-Calorimeter model (dds-CP400) was used to measure the heat created by a sample burned under an oxygen atmosphere in a closed vessel, which is surrounded by water, under controlled conditions. The measurement result is called the calorific value (MJ/kg), which is a measure of the heat energy content.

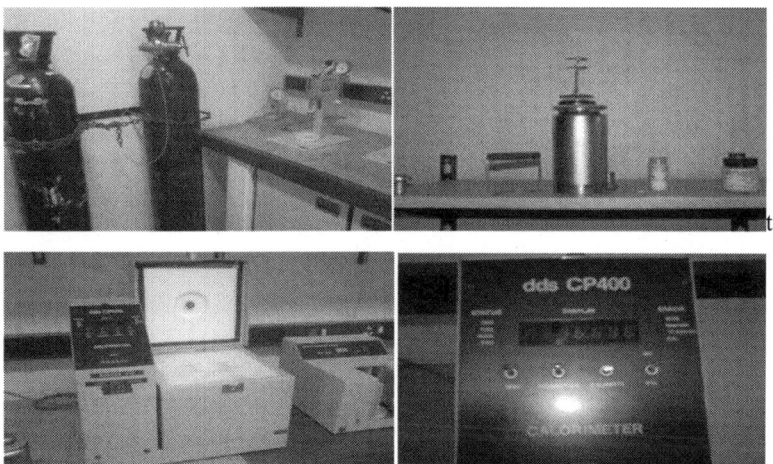

Figure 6.2: A laboratory bomb calorimeter used to measure the calorific value of the eco-fuel briquettes (Pictures taken at University of Johannesburg physical Metallurgy laboratory).

A dry solid sample was ground to using a pestle and mortar, approximately 5g of the sample was weighed into a crucible and inserted into the titanium cup shown in top right of figure 6.2. The measured sample mass was entered into the calorimeter to three decimal place. An electrical wire was carefully inserted into the crucible, ensuring that there is electrical contact between the sample and the wire. The titanium cup was filled with technical oxygen at 4,000kPa and cooled before it was inserted into the calorimeter. The calorimeter was started, and the system automatically ignites the mixture until is completely burned and measures the energy released per mass of the sample burned. At the end of the measurement, the calorific value is displayed on the unit as shown in bottom right of figure 6.2.

6.1.3 Measurement of flue gas emissions

Measurement of the flue gas emission resulting from combustion of eco-fuel briquettes was measured using Testo 350s emission analyser. The analyser is coupled with a multifunction probe which periodically samples the gas and analyse it through a series of built-in electrochemical cells. The results are recorded into the device memory which can be exported into an excel spread sheet, the results are reported in volume percent or parts per million (ppm).

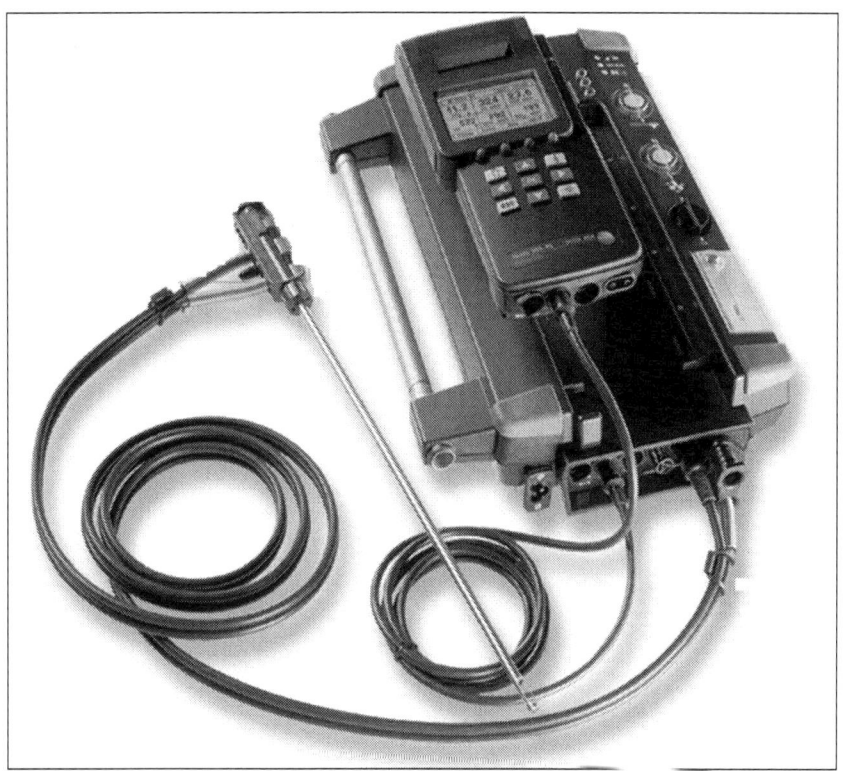

Figure 6.3: Testo 350s - emission testing and combustion analysis used for emissions analysis for eco-fuel briquettes (http://www.testo.com/online/abaxx-part.htm)

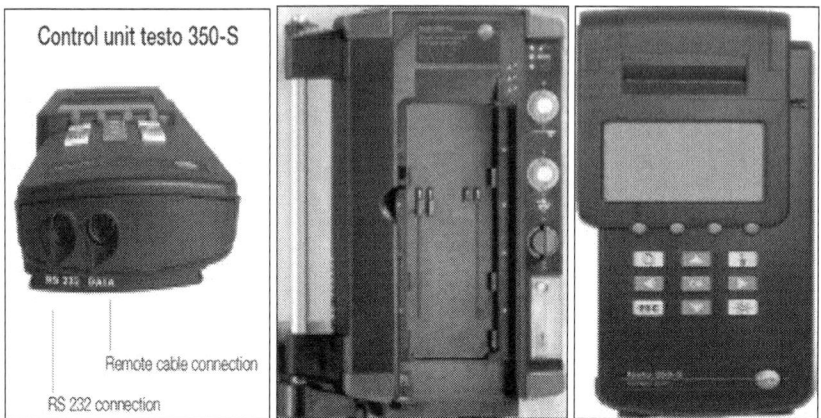

Figure 6.4: Testo 350s - emission testing and combustion analysis used for emissions analysis for eco-fuel briquettes (http://www.testo.com/online/abaxx-part.htm).

The Testo 350s emission analyzer is the most accurate combustion emissions analyzer equipment available in the Department of Geography, Environmental management and Energy studies at the University of Johannesburg Kingsway Campus. This equipment provides compliance level of accuracy, extreme testing flexibility, and the highest performance. Historically the testo 350 is chosen by more professionals for emission testing and process monitoring. The 350s helps meet the ever increasing demands of regulations and the needs of process efficiency and optimization. Exclusive sensor design, patented sample gas paths, active sample conditioning all come together for a perfect, lightweight, simple-to-use emission monitoring solution.

	O₂	CO	CO₂ₘₐₓ	NO	NOₗₒᵥ	NO₂	SO₂	H₂S	CₓHᵧ
Range	0 to 25% vol.	0 to 10,000 ppm H₂ comp.	0 to 500 ppm H₂ comp.	0 to 3,000 ppm	0 to 300 ppm	0 to 500 ppm	0 to 5,000 ppm	0 to 300 ppm	0.01 to 4%
Accuracy	< 0.2% of m.v.	< 5 ppm 0 to 99 ppm < 5% of m.v. 100 to 2,000 ppm < 10% of m.v. 2,001 to 10,000 ppm	< 2 ppm 0 to 39.9 ppm < 5% of m.v. 40 to 500 ppm	< 5 ppm 0 to 99 ppm < 5% of m.v. 100 to 2,000 ppm < 10% of m.v. 2,001 to 3,000 ppm	< 2 ppm 0 to 39.9 ppm < 5% of m.v. 100 to 300 ppm	< 5 ppm 0 to 99 ppm < 5% of m.v. 300 ppm	< 5 ppm 0 to 99 ppm < 5% of m.v. 100 to 2,000 ppm < 10% of m.v. 2,001 to 5,000 ppm	< 2 ppm 0 to 39.9 ppm < 5% of m.v. 40 to 300 ppm	< 400 ppm 100 to 4,000 ppm < 10% of m.v. > 4,000 ppm
Resolution	0.1 vol. %	1 ppm	0.1 ppm	1 ppm	0.1 ppm	0.1 ppm	1 ppm	0.1 ppm	0.001 vol. % < 10 ppm
Resp. Time	20 s (t95)	40 s (t90)	40 s (t90)	30 s (t90)	30 s (t90)	40 s (t90)	30 s (t90)	35 s (t90)	40 s (t90)

	CO₂	CO₂ Calculated	Differential Pressure 1	Differential Pressure 2	Efficiency	Flow Velocity	Current Voltage	RPM	Temperature
Range	0 to 50% vol.	0 - CO₂ max vol. %	±80" H₂O	±16" H₂O	0 to 100%	0 to 7900 ft/min	0 to 20 mA 0 to 10 V	20 to 20,000 rpm	-40 to 2192°F
Accuracy	±0.3% vol. +1% of m.v. (0 to 25% vol.) ±0.5% vol. +1.5% of m.v. (> 25 to 50% vol.)	Calculated from O₂	< 1% m.v. -20" to -80" H₂O < 1% m.v. +20" to +80" H₂O < 0.5% -19" to +19" H₂O	< 1% m.v. -16" to 1.2" H₂O < 1% m.v. +16" to +1.2" H₂O < 0.5% -1.2" to +1.2" H₂O			±0.04 mA ±0.01 V		< 33°F -40 to +212°F < 0.5% m.v. +212 to +2192°F
Resolution	0.01% vol. (0 to 25% vol.) 0.01% vol. (> 25% vol.)	0.01 vol. %	0.01" H₂O	0.01" H₂O	0.1%	10 ft/min	±0.01 mA ±0.01 V	1 ppm	

Table 6.1: Testo 350s – Technical data (http://www.testo.com/online/abaxx-part.htm).

6.1.4 Air-fuel Ratio

Air-fuel ratio (AFR) is the mass ratio of air to fuel present during combustion. When all the fuel is combined with all the free oxygen, the mixture is chemically balanced and this AFR is called the stoichiometric mixture. AFR is an important measure for anti-pollution and combustion performance. The Air fuel ratio is the most common reference term used for mixtures in internal combustion engines. It is the ratio between the *mass* of air and the mass of fuel in the fuel-air mix at any given moment (Foust, Wenzel, Clump, Maus, and Andersen, 1980).

6.2 Experimental

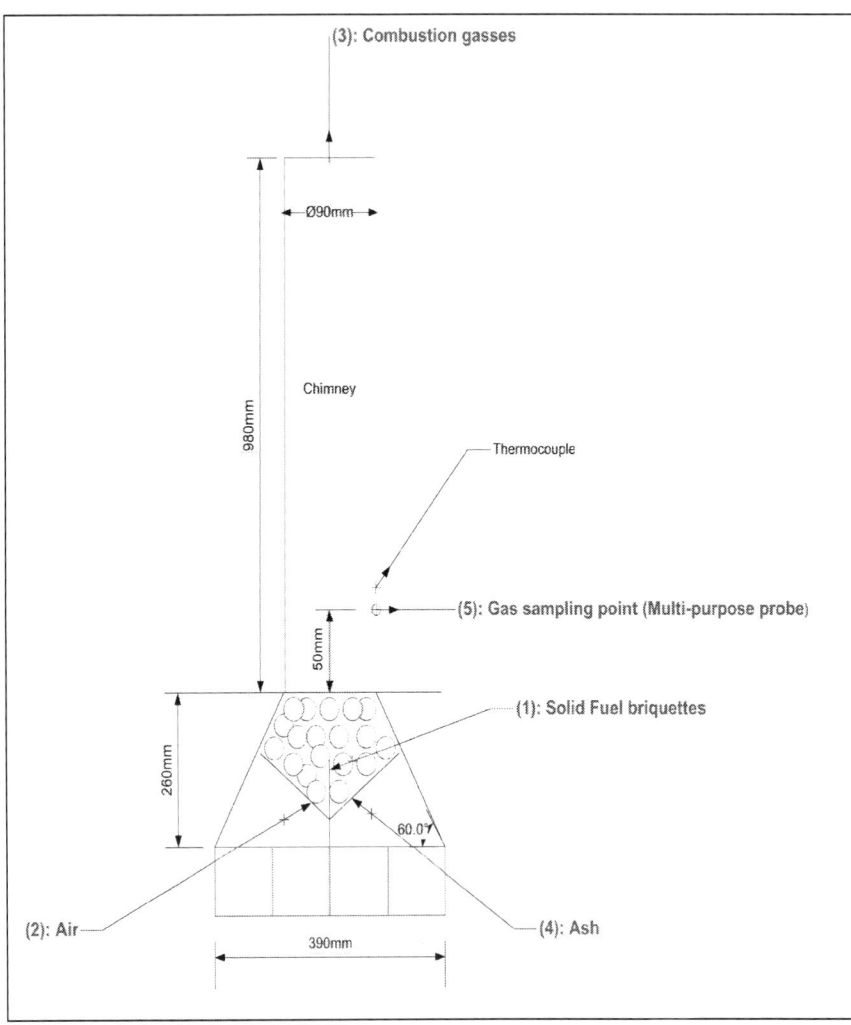

Figure 6.5: Experimental setup for gas emissions tests resulting from eco-fuel briquettes combustion.

Figure 6.6: Experimental setup for gas emissions tests resulting from eco-fuel briquettes combustion (Picture taken at University of Johannesburg FADA Centre).

The apparatus comprised of a digital scale, POCA ceramic stove, and extraction system complete with chimney. The gas extraction manifold was linked to a 980 mm stainless steel pipe of 90mm diameter chimney. A multi-purpose probe was installed 50 mm from the manifold for continuous sampling of the gas.

6.2.1 Gases

The emissions of the gaseous pollutants carbon dioxide (CO_2), carbon monoxide (CO), nitrogen oxides (NO_x), water vapour (H_2O) and sulphur dioxide (SO_2) were measured using a Testo 350s emissions analyser in real time from undiluted gas sampled directly from the chimney tunnel. Air was withdrawn via the testo multipurpose probe, which is analysed through a set of built in electrochemical cells.

6.2.2 Mass balance

Five briquettes weighing a total of 524g were put in a POCA ceramic stove together with fire wood as shown in figure. The eco-fuel briquettes were set alight using a fire lighter, they were allowed 5 minutes to stop smoking. The stove was transferred to the scale as shown in figure 6.6.A 4 litre stainless steel pot was filled with cold water and put on to the stove. Data on the heat burning characteristics, heat transfer and gas emissions were capture by the Testo 350s device and exported to an excel spread sheet for analysis.

Figure 6.7: Pictorial representation of gas emission testing (Pictures taken at University of Johannesburg FADA Centre).

6.3 Results and discussions

6.3.1 Combustion gas quality

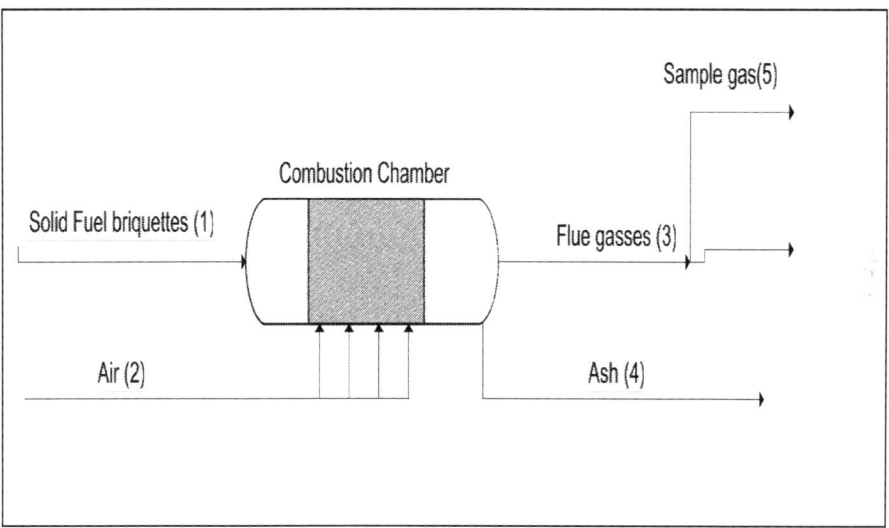

Fig: 6.8: Process flow diagram for the experimental setup used for eco-fuel briquettes emissions testing.

A measured quantity of dry eco-fuel briquettes was burnt in a POCA ceramic stove (combustion chamber). The velocity of natural incoming air was measured using an anemometer; the cross-sectional area of the air incoming path was also calculated in order to estimate the flow rate of the air. The combustion gasses were periodically sampled at point 5 as indicated in figure 6.5 & 6.6 and these were analysed through Testo 350s emission analyser. A summary of the flue gas quality is presented in table 6.2.

	O_2	CO_2	CO	H_2	H_2O	H_2S	SO_2	NO	NO_2	N_2
C_x	12.80vol%	21.33vol%	73.78ppm	59.80ppm	55,519ppm	4.32ppm	3.67ppm	1.34ppm	2.73ppm	60.30vol%
$m_x(g)$	144.99	293.32	0.067	0.035	31.52	0.005	0.006	0.001	0.004	584.77
$n_x(mol)$	4.531	6.666	0.002	0.035	1.751	0.00014	0.00012	0.00004	0.00009	20.885
mol, Ratio	104,057	153,104	55	797	40,216	3	3	1	2	479,645
$mol\%$	13.3770%	19.6821%	0.0071%	0.1024%	5.1699%	0.00004%	0.00004%	0.00001%	0.00003%	61.6603%
$mass\%$	13.747%	27.811%	0.006%	0.003%	2.988%	0.000%	0.001%	0.000%	0.000%	55.443%

Table 6.2: Summary of the gases resulting from combustion of eco-fuel briquettes

The gas produced from the combustion of eco-fuel briquettes consisted of various number of gas components such as carbon dioxide (CO_2), carbon monoxide (CO), hydrogen sulphide (H_2S), nitric oxide (NO), sulphur dioxide (SO_2), water vapour (H_2O), hydrogen (H_2), nitrogen (N_2) and oxygen (O_2). The gas was mostly dominated by inert atmospheric nitrogen, oxygen, carbon dioxide and water vapour. Figures 6.9(a) and 6.9(b) show the significance of each gas in the stream relative to each other.

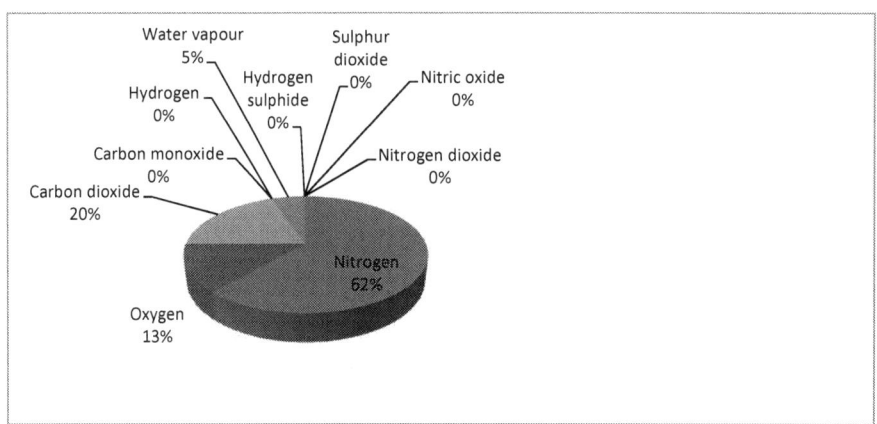

Figure 6.9(a): Representation of flue gas quality in volume %

The smaller concentrations of nitrogen and sulphur containing gases in the flue is justified by the elemental composition of the solid fuel briquette having less that 1.5 mass percent of both nitrogen and sulphur as indicated in table 6.3.

Ultimate analysis: Product						
	Carbon	Hydrogen	Oxygen	Nitrogen	Sulphur	Calorific Value
Eco-fuel briquette	36.65%	4.60%	36.30%	0.75%	0.34%	18.9 MJ/kg
Proximate Analysis: Product						
	Ash	Moisture	Fixed Carbon	Volatile Matter	Density	
Eco-fuel briquette	10.46%	10.90%	26.30%	39.34%	721kg/m^3	

Table 6.3: Fuel briquette elemental composition measured using ICP-Optima model 2100DV

Eco- fuel briquettes are the cleanest solid fuels as compared to other fossil fuels. Composed primarily of 36.65% Carbon, 36.30% Oxygen, 4.6 % Hydrogen, 0.75% Nitrogen, and 0.34% Sulphur and 10.46% Ash as indicated in table 6.3. The main products of the combustion of briquettes are nitrogen, oxygen, carbon dioxide and water vapour, the same compounds we exhale when we breathe. Wood, Coal and oil are composed of much more complex molecules, with a higher carbon ratio and higher nitrogen and sulphur contents. This means that when combusted, wood, coal and oil release higher levels of harmful emissions, including a higher ratio of carbon emissions, nitrogen oxides, and sulphur dioxide.

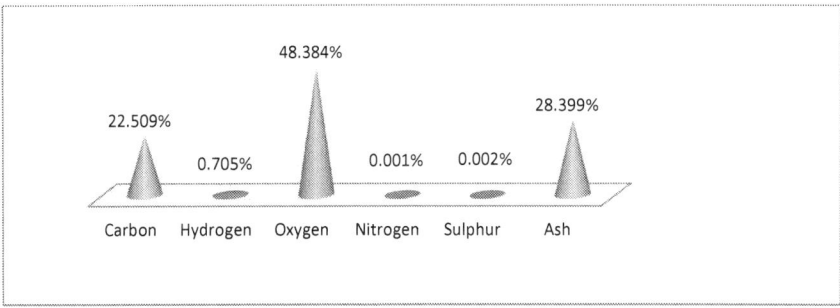

Figure 6.9(b): Calculated eco-fuel briquette composition based on the gas analysis

The calculated carbon, hydrogen, sulphur and nitrogen contents in the fuel are lower than the measured composition; this is due to the fact that the combustion took place at lower temperatures and the actual combustion process was not 100% complete. The higher oxygen content is due to excess atmospheric air not taken into consideration. The measured ash content of 10.46% is much lower than the one obtained during combustion tests. This higher ash was a result of incomplete combustion and lower temperatures.

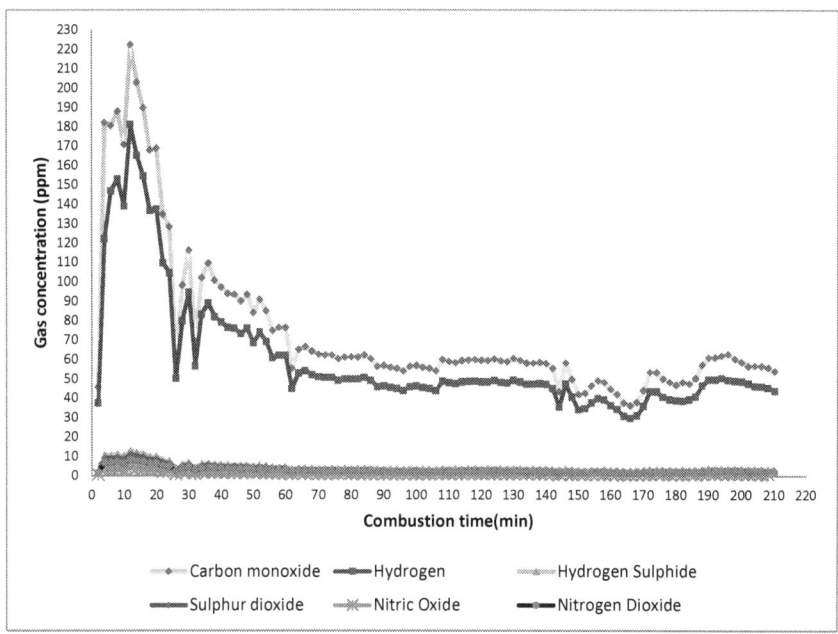

Figure 6.10: Low concentration gases in the flue gas.

Figure 6.10 shows the variations of low concentration flue gasses with time. There are high concentrations of carbon monoxide and hydrogen gas during the first 30 minutes of combustion process; however the concentrations become more stable after 60 minutes.

6.3.2 Energy balance in combustion

The following approach is used to perform energy balance during combustion based on the test results obtained. The following section covers the basic material and energy balance calculation based on the procedure stipulated by Himmelblau

(1996). All the symbols and units used in this section are fully defined in section 6.6.

The net calorific value of the briquette sample was measured to be 18,900kJ/kg.

$$800C_{3.47875}H_{1.8625}O_{6.085} + 2,3165O_2 + 8,714N_2 \rightarrow 2,782CO_2 + CO + 731H_2O + 14H_2 + 1,890O_2$$

$$+8,714N_2 \dots\dots\dots\dots\dots\dots\dots\dots\dots\dots\dots\dots\dots\dots\dots\dots\dots\dots\dots\dots\dots\dots\dots\dots\dots\dots\dots\dots\dots\dots\dots\dots\dots\dots6.1$$

$$\left[\Delta H_{rxn} = -18,900kJ.kg^{-1}\right]$$

According to Hess's Law

$$\Delta H = \left[\sum_{products}(\Delta H^0{}_f)\right] - \left[\sum_{Reactants}(\Delta H^0{}_f)\right] \dots\dots\dots\dots\dots\dots\dots\dots\dots\dots\dots\dots\dots\dots\dots\dots\dots\dots\dots6.2$$

The standard enthalpies of formation are for the relevant compounds and elements are given in table 6.4

Substance	$\Delta H^0{}_f (kJ/mol)$
$CO_2(g)$	−393.52
$CO(g)$	−110.53
$H_2O(g)$	−241.83
$H_2S(g)$	−20.63
$NO_2(g)$	+33.10
$NO(g)$	+90.29
$SO_2(g)$	−296.84
$N_2(g)$	0.00
$O_2(g)$	0.00

Table 6.4: Standard enthalpies of formation at 25 ^0C and 1atm (source: Perry & Green, 1998).

Assumption; Pressure effects are negligible

$i.e. P_0 = 82.9 kPa ; T_0 = 25^0 C$

6.3.2.1 Enthalpy calculations

<u>Enthalpy of products</u>

$$\sum_{products} \left(\Delta H^0{}_f \right) = \left(\begin{array}{l} 2,782 .\Delta H_f{}^0{}_{CO_2} + 1.\Delta H_f{}^0{}_{CO} + 731 .\Delta H_f{}^0{}_{H_2O} + 14 .\Delta H_f{}^0{}_{H_2} + 1,890 .\Delta H_f{}^0{}_{O_2} \\ + 8,714 .\Delta H_f{}^0{}_{N_2} \end{array} \right) kJ / mol$$

$$\therefore \sum_{products} \left(\Delta H^0{}_f \right) = \left(2,782 .(-393 .52) + 1.(-110 .525) + 731 .(-241 .83) + 14 .(0) + 1,890 .(0) \right) kJ / mol$$

$$\therefore \Delta H^0{}_{f_{products}} = -1.285576 \ GJ \ / \ mol$$

Pure elements at standard conditions have zero enthalpy of formation; therefore the total enthalpy of formation for the products is computed as follows:

<u>Enthalpy of reactants</u>

$$\sum_{reac\ tan\ ts} \left(\Delta H^0{}_f \right) = \left(\Delta H_f{}^0{}_{fuel} + \Delta H_f{}^0{}_{N_2} + \Delta H_f{}^0{}_{O_2} \right) kJ \ / \ mol$$

$$\sum_{reac\ tan\ ts} \left(\Delta H^0{}_f \right) = \left(800 .\Delta H_f{}^0{}_{fuel} + 8,714 .(0) + 2,316 \ (.0) \right) kJ \ / \ mol$$

$$\therefore \Delta H^0{}_{f_{reac\ tan\ ts}} = 800 .\Delta H_f{}^0{}_{fuel} \ kJ \ / \ mol$$

$but \ \Delta H_{rx} = \Delta H_f^0 \ _{products} - \Delta H_f^0 \ _{reac \ tan \ ts}$

$i.e.\ 18,900 \ kJ.kg^{-1} = -1.28566 \ GJ \ / \ mol - \left(800.\Delta H_f^0 \right)_{fuel}$

$\therefore \left(\Delta H_f^0 \right)_{fuel} = -\dfrac{\left(18,900 \ kJ.kg^{-1} \times \dfrac{423 \times 10^{-3} \ kg}{3 \ mol}\right) + \left(1.285567 \times 10^6 \ kJ.mol^{-1}\right)}{800}$

$\therefore \left(\Delta H_f^0 \right)_{fuel} = -1,610.3 \ kJ.mol^{-1}$

Total Enthalpy Calculations

Reactants

The average heat capacity of fuel is estimated based on the fuel composition and heat capacities of individual elements obtained from the literature at 298.15K.

Component	Mass %	C_{p-i}
C	36.65%	$709J.kg^{-1}.K^{-1}$
H	4.6%	$28,769J.kg^{-1}.K^{-1}$
N	0.75%	$1,042J.kg^{-1}.K^{-1}$
O	36.30%	$920J.kg^{-1}.K^{-1}$
Ash	10.46%	$0.8J.kg^{-1}.K^{-1}$
		$1,141.48J.kg^{-1}.K^{-1}$

Table 6.5: Chemical composition of fuel specific heat capacity values obtained from Perry & Green (1998).

$$H_{reac\,tan\,ts} = n_i \left(\sum_i^n \Delta H_f^0 i ....n \right) + n_i \sum_i^n \left(\int_{t_0}^t C_p .i.dt ....n \right) .... 6.3$$

The reactants are based at reference temperature and pressure.

$$H_{reac\,tan\,ts} = n_{1_{fuel}} \left(\Delta H_f^0{}_{fuel} + 1.141\,kJ.kg^{-1}.K^{-1}(0) \right) + n_{2(O_2)} \left(\Delta H^0{}_{f(O_2)} + \int_{298.15}^{298.15} 6.5 + 1\times 10^{-3}.T.dT \right)$$

$$n_{2(N_2)} \left(\Delta H^0{}_{f(N_2)} + \int_{298.15}^{298.15} 8.27 + 0.000258\,T - 1.877 \times 10^5 T^{-2} dT \right)$$

$$H_{reac\,tan\,ts} = 3mol \left(-1,610\,kJ.mol^{-1} + 0 \right) + 4.531\,mol \left(0kJ.mol^{-1} + 0kJ.mol^{-1} \right)$$
$$+ 20.88\,mol \left(0kJ.mol^{-1} + 0kJ.mol^{-1} \right)$$

$$\therefore H_{reac\,tan\,ts} = -4,830\,kJ$$

Products

$$H_{products} = n_i \left(\sum_i^n \Delta H_f^0 i ....n \right) + n_i \sum_i^n \left(\int_{t_0}^t C_p .i.dt ....n \right)$$

The products are evaluated at flame temperature relative to the reference temperature.

Total enthalpies of gas components are computed as follows:

$$\cdot\bullet\, H_{CO_2} = n_{3(CO_2)} \left(\Delta H^0{}_{f(CO_2)} + 4.184 \int_{298.15}^{413.45} (10.34 + 0.0027\,T - 195500\,T^{-2}) dT \right)$$

$$\therefore H_{CO_2} = 6.666\,mol \left(-393.52\,kJ.mol^{-1} + 4.69\,kJ.mol^{-1} \right)$$

$$\therefore H_{CO_2} = -2,591.9\,kJ$$

$$\bullet\,.H_{CO} = n_{3(CO)} \left(\Delta H^0{}_{f(CO)} + 4.184 \int_{298.15}^{413.45} (5.34 + 0.011\,T) dT \right)$$

$$\therefore H_{CO} = 0.002\,mol \left(-110.53\,kJ.mol^{-1} + 4.46\,kJ.mol^{-1} \right)$$

$$\therefore H_{CO} = -0.2121\,kJ$$

- $H_{H_2O} = n_{3(H_2O)} \left(\Delta H^0{}_{f(H_2O)} + 4.184 \int\limits_{298.15}^{413.45} \left(8.22 + 0.000156\ T + 0.00000134\ T^2 \right) dT \right)$

$\therefore H_{H_2O} = 1.751\,mol \left(-241.83\,kJ.mol^{-1} + 4.074\ kJ.mol^{-1} \right)$

$\therefore \underline{\underline{H_{H_2O} = -416.31\,kJ}}$

- $H_{H_2} = n_{3(H_2)} \left(\Delta H^0{}_{f(H_2)} + 28.769\,kJ.kg^{-1} \right)$

$\therefore H_{H_2} = 0.035 \times 10^{-3}\,kg \left(0\,kJ.kg^{-1} + 28.769\,kJ.kg^{-1} \right)$

$\therefore \underline{\underline{H_{H_2} = 0.001\,kJ}}$

- $H_{N_2} = n_{3(N_2)} \left(\Delta H^0{}_{f(N_2)} + 4.184 \int\limits_{298.15}^{413.45} \left(6.5 + 0.00100\ T \right) dT \right)$

$\therefore H_{N_2} = 20.88\,mol \left(0\,kJ.mol^{-1} + 3.31\,kJ.mol^{-1} \right)$

$\therefore \underline{\underline{H_{N_2} = +69.11\,kJ}}$

- $H_{O_2} = n_{3(O_2)} \left(\Delta H^0{}_{f(O_2)} + 4.184 \int\limits_{298.15}^{413.45} \left(8.27 + 0.000258\ T - 187700\ T^{-2} \right) dT \right)$

$\therefore H_{O_2} = 4.531\,mol \left(0\,kJ.mol^{-1} + 3.30\,kJ.mol^{-1} \right)$

$\therefore \underline{\underline{H_{O_2} = +14.95\,kJ}}$

- $H_{Ash} = m_4 \left(\Delta H^0{}_{f(ash)} + \int\limits_{298.15}^{413.45} \left(0.8 \right) dT \right)$

$\therefore H_{Ash} = 100.97 \times 10^{-3}\,kg \left(0.1046 \times 21,377\ kJ.kg^{-1} + 0.0922\ kJ.kg^{-1} \right)$

$\therefore \underline{\underline{H_{Ash} = 494.3\,kJ}}$

$$H_{products} = \left(H_{CO_2} + H_{CO} + H_{H_2O} + H_{H_2} + H_{N_2} + H_{O_2} + H_{ash}\right)$$
$$\therefore H_{products} = \left(-2,591.9 - 0.2121 - 416.31 + 0.001 + 69.11 + 14.95 + 494.3\right)kJ$$

$$\therefore H_{products} = -2,430.06\,kJ$$

Insulated system

$$q = \Delta H + \left(\frac{u_3^2 - u_\infty^2}{2}\right) + g\Delta h$$

$$q = \left(H_{products} - H_{reac\,tan\,ts}\right) + m_3\left(\left(\frac{u_3^2 - u_\infty^2}{2}\right) + g\Delta h\right)$$

$$q = \left(-2,430.06\,kJ - 4,830\,kJ\right) + 1.15569\,kg\left(\left(\frac{\left(0.017\,m.s^{-1}\right)^2 - \left(0.0503\,m.s^{-1}\right)^2}{2}\right) + 9.81\,.m.s^{-1} \times 1.240\,m\right)$$

$$q = -7,260.06\,kJ + 14.05 \times 10^{-3}\,kJ\,\left(Cooling\right)$$
$$\therefore q = -7,260\,.kJ\,\left(Cooling\right)$$

Calculations for Heat losses

	MW (g/mol)	μ (kg/m s)	K (W/m K)	ρ (kg/m³)	Cp (J/kg K)	Pr	Re$_L$	Nu$_L$	h$_m$ (W/m² K)	Q$_{cond}$ (W)	Q$_{out}$ (W)
Air	28.8	2.38E-05	3.42E-02	0.870	1,020.00	0.7107	605.8	7.293	0.25450	4.237996	323.958671
CO$_2$	44.0	2.02E-05	2.57E-02	1.234	852.52	0.6710	1,012.4	9.248	0.24252	4.038603	323.958671
CO	28.0	2.26E-05	3.34E-02	0.781	1,030.00	0.6968	573.3	7.047	0.24018	3.999613	323.958671
H$_2$	2.0	1.10E-05	1.68E-02	0.089	14,320.00	9.3921	134.0	8.110	0.13902	2.315082	323.958671
H$_2$O	18.0	1.26E-05	2.58E-02	0.804	1,010.00	0.4933	1,058.9	8.536	0.22472	3.742188	323.958671
H$_2$S	34.0	1.38E-05	2.23E-02	0.951	843.00	0.5234	1,151.5	9.079	0.20660	3.440429	323.958671
SO$_2$	48.0	1.79E-05	1.48E-02	1.527	646.00	0.7824	1,413.5	11.502	0.17370	2.892595	323.958671
NO	30.0	1.65E-05	3.42E-02	0.837	1,005.00	0.4863	839.2	7.563	0.26393	4.395144	323.958671
NO$_2$	46.0	1.48E-05	2.72E-02	1.870	1,020.00	0.5567	2,090.3	12.487	0.34657	5.771303	323.958671
O$_2$	32.0	2.63E-05	3.47E-02	0.892	928.00	0.7045	561.7	7.002	0.24792	4.128472	323.958671
N$_2$	28.0	2.26E-05	3.36E-02	0.781	1,050.00	0.7047	574.4	7.081	0.24277	4.042793	323.958671
	31.14	2.21E-05	3.18E-02	0.89	1006.33	0.70	683.52	7.57	0.24	4.04	323.96
											327.995

Table 6.6: Physical properties of the combustion gas mixture

Physical properties of the gas mixture were evaluated at film temperature in order to estimate the internal convective heat transfer coefficient.

The gas mixture was evaluated at the following conditions:

$$\overline{T}_f = \left(\frac{140.1 + 137.1}{2} \right) = 138.6^0 C; \text{gas mixture}$$

$$\rho_g = 0.86507\, kg.m^{-3}$$

$$\mu = 2.15 \times 10^{-5}\, kg.m^{-1}.s^{-1}$$

$$k_f = 3.17 \times 10^{-2}\, W.m^{-1.0}C^{-1}$$

$$Pr = \left(\frac{\mu_g \times C_p}{k_g} \right) = \frac{2.21 \times 10^{-5}\, kg.m^{-1}.s^{-1} \times 1006.33\, J.kg^{-1}.K^{-1}}{3.18 \times 10^{-2}\, J.s^{-1}.kg^{-1}.K^{-1}} = 0.700$$

Reynolds Number

$$Re_D = \left(\frac{0.89 kg.m^{-3} \times 0.017 m.s^{-1} \times 90 \times 10^{-3}\, m}{2.21 \times 10^{-5}\, kg.m^{-1}.s^{-1}} \right) = 62 \text{ laminar flow!}$$

The gas flow inside the chimney is laminar; the following empirical correlation is satisfied by the above given conditions, assuming circular flow correlation.

Nusselt Number

$$Nu_D = C.Re_D{}^m .Pr^{\frac{1}{3}}$$

Valid for : T_f, $0.4 \le Re_D \le 4 \times 10^5$ and $Pr \ge 0.7$

for $40 \le Re_D \le 4000 : C = 0.683; m = 0.466$

$$\therefore Nu_D = 0.683.Re_D{}^{0.466} .Pr^{\frac{1}{3}}$$

$$\therefore Nu_D = 0.683.(62)^{0.466}.(0.7)^{\frac{1}{3}}$$

$$\therefore Nu_D = 3.82$$

Convective heat transfer coefficients

$2W.m^{-2}.K^{-1} \leq h_i \leq 25W.m^{-2}.K^{-1}$ (Incropera & Dewitt, 1934)

assume $h_i = 25W.m^{-2}.K^{-1}$

$h_0 = 3.63W.m^{-2}.K^{-1}$

Heat loss calculations

Heat loss by convection outside the chimney

$q_{convection} = h_o A(T_s - T_{ambt})$

$q_{convection} = 3.63\,W.m^{-2}.K^{-1}.\pi \times 90 \times 10^{-3}\,m \times 980 \times 10^{-3}\,m\,(137.1^0C - 28.3^0C)$

$\therefore q_{convection} = \underline{109.4W}$

Heat loss through flue gasses

$q_{gas} = m_{gas}.(C_p)_{gas}.(T_s - T_{ambt})$

$q_{gas} = \rho_{gas}.Q_{gas}.(C_p)_{gas}.(T_s - T_{ambt})$

$q_{gas} = 0.89\,kg.m^{-3}.1.08 \times 10^{-4}\,m^{-3}.s^{-1}.(1006.3\,J.kg^{-1}.K^{-1})_{gas}.(137.1^0C - 28.3^0)$

$q_{gas} = \underline{10.5W}$

Heat loss by radiation outside the chimney

$q_{radiation} = \varepsilon\delta A(T_s^4 - T_{ambt}^4)$

$q_{radiation} = 1 \times 5.67 \times 10^{-8}\,W.m^{-2}.K^{-4}\left(\pi.90 \times 10^{-3}\,m.980 \times 10^{-3}\,m\right) \times \left((410.25)^4 - (301.45)^4\right)$

$\therefore q_{radiation} = \underline{315.3W}$

Total heat loss

$q_{loss} = q_{covection} + q_{gas} + q_{radiation}$

$\therefore q_{loss} = \underline{109.4W + 10.5W + 315.3W}$

$\therefore q_{loss} = \underline{435.2W}$

Overall energy balance

Heat supplied = Heat absorbed by water + Heat loss to the surroundings

$$m_b \times CV = q_w + q_{loss}$$

$$\therefore q_w = \left(\frac{0.423 \ kg \times 18.9 \times 10^{6} \ J.kg^{-1}}{12,600 \ s} \right) - 435.2W$$

$$\therefore q_w = \underline{199.3W}$$

Accuracy check

$$q_w = m_{w_0} \left(C_p \right)_w .(T_b - T_o) + \left(m_{w_0} - m_{w_1} \right) \lambda$$

$$q_w = 3.792 \ kg \left(4.18 \ kJ.kg^{-1}.K^{-1} \right) \left(94.5^{\circ}C - 21.4^{\circ}C \right) + \left(3.792 \ kg - 2.800 \ kg \right) \times 2216 \ kJ.kg^{-1}$$

$$\therefore q_w = \underline{266.4W} \ (compares \quad with \ the \quad one \ calculated \quad above)$$

$$\vartheta = \frac{199.3W}{\left(\dfrac{0.423 \ kg \times 18.9 \times 10^{6} \ J.kg^{-1}}{12,600 \ s} \right)} \times 100 \ \% = \underline{31.4\%}$$

The above calculations only serve as an initial indication. For example, the flue gas temperatures vary along the height of the chimney and this could affect the radiation and convection heat transfer calculations. The gas velocity inside the chimney is mainly dependant on the ambient combustion and may also vary at any point in the chimney. It is recommended that further work is conducted in order to model temperature profile along the height of the findings.

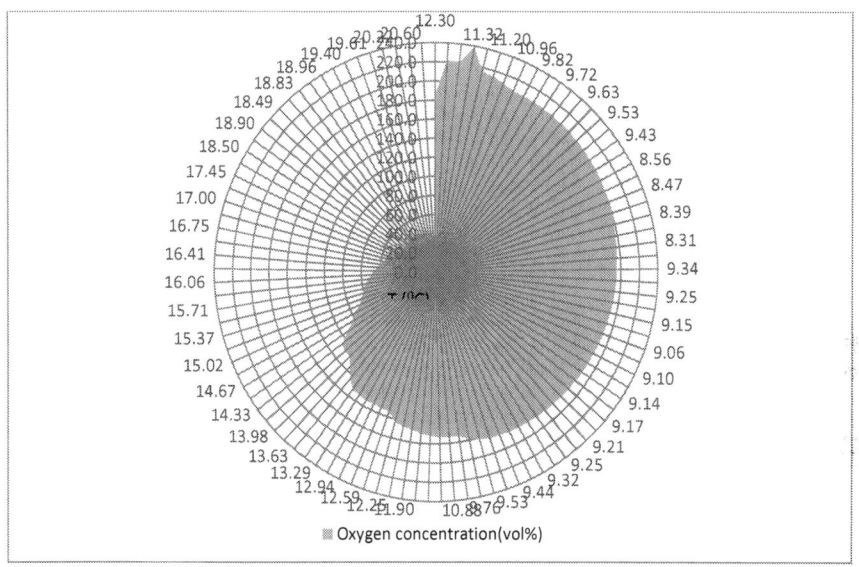

Figure 6.11: Radar representation of flame temperature with oxygen concentration in the gas stream.

The above figure shows the variation of flame temperature with the concentration of oxygen in the combustion gases. Oxygen concentration drops below 8.31% at the maximum combustion rate at a flame temperature of 197^0 C. As the flame temperature drops, the oxygen concentration in the flue gasses increases, meaning that less oxygen is consumed when the combustion process slows down. The final oxygen concentration of 20.57% is attained when the flue gas temperature is equal to the ambient temperature, this is evidence that the combustion process stopped and oxygen is no longer consumed.

6.4 Conclusions

Combustion of eco-fuel briquettes in a laboratory-scale POCA ceramic stove was investigated to evaluate its combustion characteristics and gas emission quality using the testo portable emission analyzer Testo 350s. The results show that high combustion efficiencies could be achieved by choosing appropriate operating conditions such as the air-to-fuel ratio. The efficiencies were between 91-95% for carbon utilisation efficiency and over 99.47% for CO to CO_2 combustion efficiency at an estimated air-to-fuel ratio of 1.44. The average burning rate of 2g/min was obtained from the test work, meaning that 1kg of eco-fuel briquettes can burn for 7hours which makes it ideal for domestic applications. The standard enthalpy of formation of the briquettes was estimated as $-1,619.3 kJmol^{-1}$ and the net heat loss to the surroundings was $364.6 Js^{-1}$ of which 10.7% was lost via free convection and balance by radiation. Some of these heat losses may be avoided by controlling the combustion conditions for example, pre-heating the air prior to combustion. 31.4% of the heat produced by the briquettes can be absorbed by a pot of water. The flue gas produced from this reaction consisted of 21.3% carbon dioxide, 0.0074% carbon monoxide, 0.000432% hydrogen sulphide, 0.000134% nitric oxide, 0.000367% sulphur dioxide, 5.5 % water vapour, 0.00598% hydrogen, 60.3 % nitrogen and 12.8 % oxygen. The gas was mostly dominated by inert atmospheric nitrogen, oxygen, carbon dioxide and water vapour. The test results have shown that an eco- fuel briquette is the cleanest solid fuel as compared to other fossil fuels composed primarily of 36.65% Carbon, 36.3% Oxygen, 4.6% Hydrogen, 0.75% Nitrogen, and 0.34% Sulphur and 10.46% Ash as per test results from ICP analysis. The effects of operating parameters on eco-fuel briquettes combustion, such as gas velocity and excess air, prcheated air, temperature and velocity may need to be investigated further in detail in order to improve the combustion efficiency of the briquettes under given conditions.

6.5 References

Coulson, J.M., Richardson, J.F. & Sinnott, R.K. 1999. *Chemical Engineering - Volume 6: An Introduction to Chemical Engineering Design 3rd ed*, Butterworth Heinemann.

Himmelblau, D.H. 1996. *Basic Principles and Calculations in Chemical Engineering*, 6th Edition, Prentice-Hall International.

Luyben, W.L. & Wenzel, L.A. 1988. *Chemical Process Analysis: Mass and Energy Balances*, Prentice-Hall.

Perry, R.H. & Green, D. 1998. *Perry's Chemical Engineer's Handbook*, 7th Edition, McGraw-Hill.

Reklaitis, G.V. 1983. *Introduction to Material and Energy Balances*, Wiley.

6.6 Nomenclature and Units

Symbol	Definition	SI units
A	Heat transfer area	m^2
C_p	Specific heat capacity	$J.kg^{-1}.K^{-1}$
d	Diameter of the chimney	m
d_b	Inner diameter of the briquette	m
t	Time	s
x_i	Ambient air absolute humidity of stream i	$g_{\cdot moisture} / kg_{\cdot dry\, .air}$
$m_{i(j)}$	Mass of component j in stream i	kg
$n_{i(j)}$	Moles of component j in stream i	$moles$
M_W	Molecular weight	$g.mol^{-1}$
T_3	Gas temperature at stream 3	^0C
T_s	Surface temperature	^0C
ΔT	Temperature difference	^0C
T_f	Film temperature	^0C
T_{ambt}	Ambient air temperature.	^0C
T_3	Gas temperature at stream 3	^0C
T_s	Surface temperature	^0C
ΔT	Temperature difference	^0C
T_f	Film temperature	^0C
v_3	Gas velocity at stream 3	$m.s^{-1}$
u_∞	Air velocity outside the chimney	$m.s^{-1}$
m_1	Total mass of stream 1	kg
m_2	Total mass of stream 2	kg
$m_{2(N_2)}$	Mass of nitrogen in stream 2	kg
$m_{2(O_2)}$	Mass of oxygen in stream 2	kg
m_3	Total mass of stream 3	kg
$m_{3(N_2)}$	Mass of nitrogen in stream 3	kg
$m_{3(O_2)}$	Mass of oxygen in stream 3	kg
$m_{3(CO_2)}$	Mass of nitrogen in stream 3	kg
$m_{3(CO)}$	Mass of oxygen in stream 3	kg
$m_{3(H_2O)}$	Mass of nitrogen in stream 3	kg
$m_{3(H_2S)}$	Mass of oxygen in stream 3	kg

$m_{3(NO)}$	Mass of nitrogen in stream 3	kg
$m_{3(NO_2)}$	Mass of oxygen in stream 3	kg
m_4	Total mass of stream 4	kg
n_1	Total moles of stream 1	$moles$
n_2	Total moles of stream 2	$moles$
$n_{2(N_2)}$	Moles of nitrogen in stream 2	$moles$
$n_{2(O_2)}$	Moles of oxygen in stream 2	$moles$
n_3	Total moles of stream 3	$moles$
$n_{3(N_2)}$	Moles of nitrogen in stream 3	$moles$
$n_{3(O_2)}$	Moles of oxygen in stream 3	$moles$
$n_{3(CO_2)}$	Moles of nitrogen in stream 3	$moles$
$n_{3(CO)}$	Moles of oxygen in stream 3	$moles$
$n_{3(H_2O)}$	Moles of nitrogen in stream 3	$moles$
$n_{3(H_2S)}$	Moles of oxygen in stream 3	$moles$
$n_{3(NO)}$	Moles of nitrogen in stream 3	$moles$
$n_{3(NO_2)}$	Moles of oxygen in stream 3	$moles$
n_4	Total moles of stream 4	$moles$
h	Convective heat transfer of gas mixture	$W.m^{-2}.K^{-1}$
h_{air}	Convective heat transfer coefficient of air	$W.m^{-2}.K^{-1}$
k_a	Thermal conductivity of air	$W.m^{-1}.K^{-1}$
k_g	Thermal conductivity of gas mixture	$W.m^{-1}.K^{-1}$
ΔH_{rxn}	Heat of reaction	$J.mol^{-1}$
$\Delta H^0{}_f$	Standard enthalpy formation	$J.mol^{-1}$
$H_{products}$	Total enthalpy of products	J
$H_{reactants}$	Total enthalpy of reactants	J
Δh	Difference in height	m
C_i	Gas concentration(parts per million)	$mg.l^{-1}$
g	Gravitational acceleration	$m.s^{-2}$
$\hat{v}_a$	Specific volume of moist air	$m^{-3}.kg^{-1}$
P_t	Absolute pressure	Pa
R	Ideal gas constant	$m^3.Pa.mol^{-1}.K^{-1}$
Q_3	Total gas volumetric flow rate at stream 3	$m^3.s^{-1}$
q	Energy transfer in the form of heat	J
C_p	Specific heat capacity	$J.kg^{-1}.K^{-1}$

CV	Calorific Value	$MJ.kg^{-1}$
Nm^3	Volume at standard pressure and temperature	m^3

λ	Latent heat of water	$kJ.kg^{-1}$
$\eta_{combustion}$	Combustion efficiency	$\%$
δ	Stefan's Boltzmann's constant	$W.m^{-2}.K^{-4}$
ρ_g	Gas density	$kg.m^{-3}$
μ_g	Viscosity of air	$kg.m^{-1}.s^{-1}$
$\%RH$	Percent relative humidity	$\%$
$\Pr$	Prandlt Number	*Dimensionless*
Re_L	Nusselt Number	*Dimensionless*
Nu_L	Reynolds Number	*Dimensionless*
ε	Emissivity	*Dimensionless*
AFR	Air fuel ratio	*Dimensionless*
y_{O_2}	Mole fraction of oxygen in the gas phase	*Dimensionless*
SG	Specific gravity	*Dimensionless*

CO_2	Carbon dioxide	gas
CO	Carbon monoxide	gas
H_2O	Water	gas
H_2S	Hydrogen Sulphide	gas
NO	Nitric oxide	gas
NO_2	Nitrogen dioxide	gas
SO_2	Sulphur dioxide	gas
N_2	Nitrogen gas	gas
O_2	Oxygen gas	gas
C	Carbon	Solid
H	Elemental Hydrogen	
O	Elemental oxygen	
N	Elemental Nitrogen	
S	Elemental Sulphur	

CHAPTER 7: Economic evaluation

7.1 Background

This section discusses the relationship between technical and commercial concepts in order to design a feasible and economically viable process. Before any detailed work is done on the design, the technical and economics of the proposed process should be examined (Allen, 1991). The various reactions and physical processes involved must be considered, along with the existing and potential market conditions for the particular product. A preliminary survey of this type gives an indication of the probable success of' the project and also shows what additional information is necessary to make a complete evaluation (Peters, & Timmerhaus, 1991).

The present value of an asset may be defined as the value of the asset in its condition at the time of valuation. There are several different types of present values, and the standard meanings of the various types should be distinguished (Garrett, 1989).

$$P_v = \frac{f_v}{(1+i)^n} \dots\dots\dots\dots\dots\dots 7.1$$

$P_v = $ Present value

$f_v = $ Future value

$i = $ Interest rate

7.1.1 Depreciation

Depreciation costs can be determined by a number of different methods, and the design engineer should understand the bases for the various methods. The Federal government has definite rules and regulations concerning the manner in which depreciation costs may be determined. These regulations must be followed for income-tax purposes, as well as to obtain most types of governmental support (Ulrich, 1984). Since the methods approved by the Government are based on sound economic procedures, most industrial concerns use one of the Government methods for determining depreciation costs, both for income-tax calculations and for reporting the concern's costs and profits (Peters & Timmerhaus, 1991). It is necessary, therefore, that the design engineer keep abreast of current changes in Governmental regulations regarding depreciation allowances. In general, depreciation accounting methods may be divided into two classes: arbitrary methods giving no consideration to interest costs, and methods taking into account interest on the investment. Straight-line, declining-balance, and sum-of-the-years-digits methods are included in the first class, while the second class includes the sinking-fund and the present-worth methods (Perry & Green, 1998).

7.1.2 Straight-Line Method

In the straight-line method for determining depreciation, it is assumed that the value of the property decreases linearly with time. Equal amounts are charged for depreciation each year throughout the entire service life of the property. The annual depreciation cost may be expressed in equation form as follows (Garrett, 1989).

$$d = \frac{V - v_s}{n} \dots\dots\dots\dots\dots\dots\dots\dots\dots\dots\dots\dots\dots\dots\dots\dots\dots\dots\dots\dots\dots\dots\dots\dots\dots\dots\dots\dots\dots\dots\dots\dots.7.2$$

Where d = annual depreciation, R/year

V = original value of the property at start of the service-life period, completely installed and ready for use.

v_s = salvage value of property at end of service life.

n = service life, years

The asset value (or book value) of the equipment at any time during the service life may be determined from the following equation:

$$V_a = V - ad \dots\dots\dots\dots\dots\dots\dots\dots\dots\dots\dots\dots\dots\dots\dots\dots\dots\dots\dots\dots\dots\dots\dots\dots\dots\dots\dots\dots\dots\dots\dots.7.3$$

Where V_a, = asset or book value; a = the number of years in actual

Because of its simplicity, the straight-line method is widely used for determining depreciation costs. In general, design engineers report economic evaluations on the basis of straight-line depreciation unless there is some specific reason for using one of the other methods (Peters & Timmerhaus, 1991).

7.1.3 Profitability

The word profitability is used as the general term for the measure of the amount of profit that can be obtained from a given situation. Profitability, therefore, is the common denominator for all business activities. Before capital is invested in a project or enterprise, it is necessary to know how much profit can be obtained and whether or not it might be more advantageous to invest the capital in another form of enterprise (Perry & Green, 1998). Thus, the determination and analysis of profits obtainable from the investment of capital and the choice of the best investment among various alternatives are major goals of an economic analysis.

There are many reasons why capital investments are made. Sometimes, the purpose is merely to supply a service which cannot possibly yield a monetary profit, such as the provision of recreation facilities for free use of employees. The profitability for this type of venture cannot be expressed on a positive numerical basis. The design engineer, however, usually deals with investments which are expected to yield a tangible profit. Because profits and costs are considered which will occur in the future, the possibilities of inflation or deflation affecting future profits and costs must be recognized (Peters, & Timmerhaus, 1991). The profitability index is defined as follows:

$$PI = \frac{NPV - I_{out}}{I_{out}} \dots\dots\dots\dots\dots\dots\dots\dots\dots\dots\dots\dots\dots\dots\dots\dots\dots\dots\dots\dots\dots\dots\dots\dots 7.4$$

A positive profitability index is attractive to investors.

Continuous interest compounding is an important concept for profitability evaluation, in most cases; the interest is generally treated as finite-period interest compounded annually. According to the relationships developed in chapter 7, Peters & Timmerhaus (1999) it is a simple matter to convert to the case of continuous interest compounding in place of finite interest compounding.

For example, the discount factor

$$d_n = \frac{1}{(1+i)^n} \dots\dots\dots\dots\dots\dots\dots\dots\dots\dots\dots\dots\dots\dots\dots\dots\dots\dots\dots\dots\dots\dots\dots\dots 7.5$$

7.1.4 Cost of Capital

The simplest approach is to assume that investment of capital is made at a rate of return equivalent to the total profit, or rate of return, over the full expected life of the particular project. This method has the advantage of putting the profitability analysis of all alternative investments on an equal basis, thereby permitting a clear comparison of risk factors. This method is particularly useful for preliminary estimates, but it may need to be refined further to take care of income-tax effects for final evaluation (Coulson, Richardson & Sinnott, 1999).

7.2 Mathematical Methods for Profitability

7.2.1 Evaluation

The most commonly used methods for profitability evaluation, as illustrated by Garrett (1989) and are listed as follows:

1. Internal Rate of Return (IRR)
2. Net Present Value (NPV)
3. Profitability Index (PI)
4. Payback Period (p_{period})

Each of these methods has its advantages and disadvantages, and much has been written on the virtue of the various methods. Because no single method is best for all situations, it is important to evaluate each method independently in order to arrive at a decision with the highest confidence level.

7.2.1.1 Internal rate of return

In engineering economic studies, rate of return on investment is ordinarily expressed on an annual percentage basis. The yearly profit divided by the total initial investment necessary represents the fractional return, and this fraction times 100 is the standard percent return on investment. Profit is defined as the difference between income and expense. Therefore, profit is a function of the quantity of goods or services produced and the selling price (Coulson, Richardson & Sinnott, 1999). The amount of profit is also affected by the economic efficiency of the operation, and increased profits can be obtained by use of effective methods which reduce operating expenses. To obtain reliable estimates of investment returns, it is necessary to make accurate predictions of profits and the required investment (Garrett, 1989). To determine the profit, estimates are made of direct production costs, fixed charges including depreciation, plant overhead costs, and general expenses. Profits may be expressed on a before-tax or after-tax basis, but the conditions should be indicated. Both working capital and fixed capital should be considered in determining the total investment.

7.2.1.2 Net Present Value

A related approach, known as the method of net present worth or net present value or venture worth, substitutes the cost of capital at an interest rate for the discounted-cash-flow rate of return (Garrett, 1989). The cost of capital can be taken as the average rate of return the company earns on its capital, or it can be designated as the minimum acceptable return for the project. The net present worth of the project is then the difference between the present value of the annual cash flows and the initial required investment.

$$NPV = PV_{in} - I_{out} \dotfill 7.6$$

7.2.1.3 Payback Period

Payback period, or payout time, is defined as the minimum length of time theoretically necessary to recover the original capital investment in the form of cash flow to the project based on total income minus all costs except depreciation. Generally, for this method, original capital investment means only the original, depreciable, fixed-capital investment, and interest effects are neglected (Garrett, 1989).

Thus,

$$P_{period} = \frac{K_f}{\text{Pr}\,ofit} \dotfill 7.7$$

Another approach to payback period takes the time value of money into consideration and is designated as payback period including interest. With this method, an appropriate interest rate is chosen representing the minimum acceptable rate of return. The annual cash flows to the project during the estimated life are discounted at the designated interest rate to permit computation of an average annual figure for profit plus depreciation, which reflects the time value of money. The time to recover the tied-capital investment plus compounded interest on the total capital investment during the estimated life by means of the average annual cash flow is the payout period including interest (Garrett, 1989).

7.3 Discussion of the Bikernmayer Model

Below is a schematic process flow diagram showing the production steps of eco-fuel briquettes.

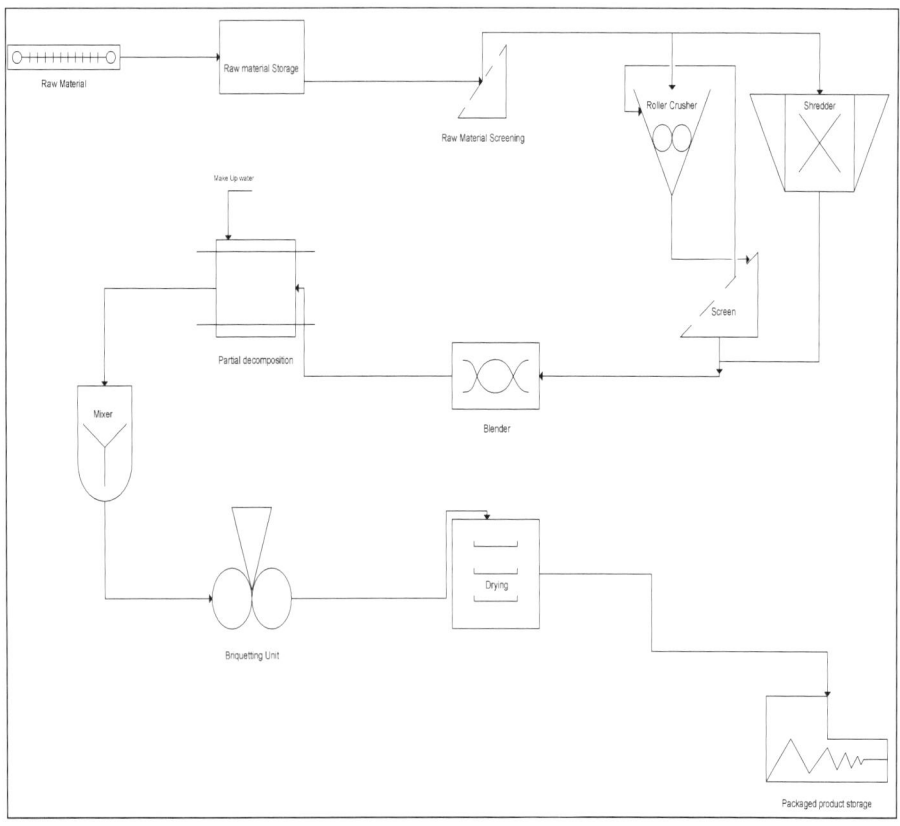

Figure 7.1: Process flow diagram for the eco-fuel briquettes production

7.3.1 The economic evaluation process

The proposed eco-fuel briquettes production line consists of four preparation stages prior to pressing and drying. The preparation stages includes, hand sorting, screening, size reduction, and mixing. The final stage of preparation involves soaking of combined material in water for a specific period of time to allow partial re-pulping and decomposition of plant fibres. The material is therefore processed using a compaction unit to form briquettes which are later dried and packed for the consumers.

The findings obtained from the technical part of this dissertation will form the basis of costing and economic considerations. To ensure accurate costing, the costs of equipment required for the product were obtained from the quotations supplied by the manufacturers and suppliers and the prices are based on the 2009 consumer price index. This section will outline a summary of the factors taken into account when developing an economic model which can be used as tool of deciding on the economic viability of this project. The steps taken follow the procedure suggested by Peters & Timmerhaus (1999).

Raw materials (availability, quantity, quality & cost)

Raw material required for this is usually biomass waste such as saw dust, wood chips, grass clippings, waste paper, mielie husks, coffee grounds and coal fines. These materials are generally regarded as waste and can be obtained at zero cost. However transportation cost should allowed for between the waste generator and the production facility a 15% handling fee has been allowed for to account for raw material gate fee, in most cases the waste generator may subsidise the transport as eco-fuel briquettes is considered an environmentally friendly method of re-utilizing certain waste material which in turn saves the waste generator on general landfill disposal fees. A study on the availability of raw material in the townships has shown a positive potential for this project. Furthermore, the availability of the raw material from production industries in large quantities and consistent quality increases the confidence of producing briquettes of consistent quality.

Thermodynamic, physical and chemical properties

Thermodynamic, chemical and physical factors affecting the production have been studied in chapter 5 and information is available to be used in developing the economic model.

Facilities and equipment available at present

The production facility Thuthukani and equipment suitable for processing 200kg per day is currently available at Tsakani in the East Rand .This facility has been used in the past for piloting and product –concept launch in May 2009.

Facilities and equipment which must be purchased

The existing land has an area of $4,655m^2$ available for expansion as per information obtained from Google Earth satellite images; the land is fully developed, with a fence, 24 hour security, water and sanitation.

Estimation of production costs and total investment

The process is designed to minimise capital investment and ensure that more jobs for unskilled individuals are created. Profits, probable and optimum, per ton of product and per year, has been estimated based on the current energy price taking in to account an indication or the realistic price which the targeted market is willing to pay. Materials of construction for the capital equipment were selected based on the durability and price.

Markets Survey

A market survey was conducted on the product and there is a potential demand of the briquettes in the township. Present and future supply and demand, present uses, new uses, present buying habits, price range for products and by-products, character, location, and number of possible customers were studied.

Competition

This is low cost product targeting people in the townships and informal settlements, based on the market survey response; there is a slight competition within the informal fuel wood trading. Based on the market survey conducted in March 2009, fuel wood is trading at R2.80/kg (17.5cents/MJ) compared to renewable eco-fuel briquettes priced at R2.26/kg (11.9cents/ MJ). However a detailed survey may be conducted in future to look in to the overall production statistics, comparison of various manufacturing processes and product specifications of competitors should they exist within the same territory.

Properties of product (Fuel briquette)

Detailed Chemical and physical properties, specifications and impurities of the product is fully discussed in chapter 6.

Sales and Marketing

Method of selling and distributing, advertising required, technical services required will be undertaken by the sales and marketing team of Phumani Paper. The cost of marketing has been allowed as part of the contingencies in the total project cost.

Patent situation and legal restrictions

Phumani paper has bought a patent license for the Porta press and the screw press was designed for Phumani paper. A patent of the Bikernmayer manual hand press is to be negotiated between Phumani Paper and Bikernmayer.

7.4 Discussions of the Bikernmayer Financial Model

The manual hand press proposed by Bikernmayer seems to be the most economical piece of equipment to be used in producing the eco-fuel briquettes. The total capital investment of this equipment is lower than that of Porta press and screw press. Furthermore, the Bikernmayer press has the shortest production cycle time and larger capacity compare to the other tested manual presses (Porta and Screw presses). The evaluation tests were based on producing the proposed hollowed cylindrical shape eco-fuel briquettes.

7.4.1 Mass Balance

The total cycle time obtained experimentally using the Bikernmayer press was 0.17 minutes for producing four briquettes per cycles, each weighing 200g. The Bikernmayer press is able to process 800g of raw material in 10.2 seconds (282.32 kg/hr). The number of briquettes processed in an hour is calculated as follows:

$$n_b = \frac{282.35 kg.hr^{-1}}{0.2kg.brick} = 1412 bricks.hr^{-1}$$

The model allows for a production schedule of 8 hour normal shift per day for 22 days per month. This result in the daily, monthly and annual production rates estimated as follows:

$$d_r = \frac{282.35kg}{hr} \times \frac{8hr}{1day} = 2,258.8kg.day^{-1}$$

$$m_r = \frac{2,258.8kg}{dat} \times \frac{22d}{1month} = 49,693.6kg.month^{-1}$$

$$a_r = \frac{49,693.6kg}{month} \times \frac{12month}{1year} = 596,332.2kg.year^{-1}(596.32tons/year)$$

The drying test work conducted in chapter 5 of this indicates that the briquettes can be dried from 52.85% solids to 89.27%solids. Based on these findings, the dry monthly production rate may be calculated as follows:

$$m_{r(dry)} = 49,693.6kg.month^{-1} \times \frac{89.27}{100\%} = 44,361.5kg/month$$

Mass of dry briquette:

$$m_{brick(dry)} = 200g \times \frac{0.5285}{0.8927} = 118.4g @ 10.73\% moisture$$

Equipment Capacities	Total Cycle time(min)	Brick(s) mass(g)	Produc. Rate(kg/h)	No.Briq/h	Prod. Rate(kg/d)	Prod. Rate(kg/m)	No.Briq/m
Screw Press(Four presess)	3.5	200	13.71	69	110	2,414	12,069
Porta Press(Four presess)	0.5	200	96.00	480	768	16,896	84,480
Birkennmeyer brick press(One Press)	0.17	800	282.35	1412	2,259	49,694	248,471

Table 7.1: Equipment capacities for the screw press, Porta Press and Bikernmayer press.

The above table shows the Bikernmayer press being the most cost effective equipment (R750/unit) compared to the Porta press (R1200/unit) and screw press (R850/unit) but yet having the largest production capacity of about 12 times the Porta press and 83 times the screw press.

Briquette composition

Raw Material	Sample Mass(kg)	% (w/w)	Screening	Shredding	Decomposing	Griding	Mixing	Collection	Tranportation	Storage	Actual Mass(kg/m)
Spent coffe ground	4.3	31.91	1	0	0	1	1	1	1	1	15,855
Finely grinded coal fines	3.1	23.04	1	0	1	1	1	1	1	1	11,451
Saw dust/wood chips	1.5	11.08	0	0	0	0	1	1	1	1	5,505
Mielie husk/grass clippings/cow dung	2.4	17.73	0	1	1	0	1	1	1	1	8,808
Granulatted Paper(news print/card board)	1.4	10.34	0	1	1	0	1	1	1	1	5,138
Pulp containing water	0.8	5.91	1	0	0	0	1	1	1	1	2,936
Total material	13.5	100									49,694

Table 7.2: The proposed fuel briquette compositions and estimated raw material requirements for the Bikernmayer model

Table 7.2 shows the composition of the proposed eco-fuel briquettes with a sample size of 13.5kg. The zero represents non-required services and represents required services. This model requires large quantities of raw material as compared to others, thus a bakkie and trailer is suitable for continuous collection raw material whereas wheel barrows and trolleys are used for the other models. The above mentioned recipe is not fixed, any combination of biomass and binding organic material may be used to produce the briquettes depending on the availability of the raw material.

Raw Material	Screw Press(kg/m)	Porta Press(kg/m)	Birk.brick press(kg/m)
Spent coffe ground	770	5,391	15,855
Finely grinded coal fines	556	3,893	11,451
Saw dust/wood chips	267	1,872	5,505
Mielie husk/grass clippings/cow dung	428	2,995	8,808
Granulatted Paper(news print/card board)	250	1,747	5,138
Pulp containing water	143	998	2,936
Total Raw mat. Requirements	2,414	16,896	49,694
Ancillaries	No. Operators	No. Operators	No. Operators
Hand operated Mixer	2	2	2
Hand operated grinder/crusher	2	2	2
Screening	1	1	2
Hand operated Press	8	8	2
Hand operater shredder	2	2	2
Stock Piling of raw material	0	0	0
Rain proof drying land	1	1	1
Water & Electricity		0	0
General Waste disposal		0	0
Product Packaging	0	0	0
Product storage	1	1	4
Raw material Collection & transportation	6	6	2
Raw material storage	1	1	1
Total number of people	24	24	18

Table 7.3: Required raw material for the screw press, Porta press and Bikernmayer press as well as the labour required for each press.

ITEM No	DESCRIPTION	UNITS	QTY			COMMENTS
1				EQUIPMENT COST		
1.1	BIRKENMEYER MANUAL HAND PRESS					
1.1.1	Design & Setup cost		1	R	5,500.00	Design & setup cost is once off.
1.1.2	Fabrication cost		1	R	750.00	Each press can produce 4 bricks/cycle(2x operators)
				R	6,250.00	Based on R750 Per Press
1.2	RAW MATERIAL HANDLING					
1.2.1	Storage tanks		12	R	1,822.60	Each tank costs R146
1.2.2	Screens		1	R	320.00	1X Operators required @ R320 each
1.2.3	Hand operated Mixer		1	R	3,200.00	500kg Capacity(2X operators)
1.2.4	Hand operated Grinder		1	R	560.00	1X Operators required
1.2.5	Hand operated Shredder		1	R	180.00	1X Operators required
	TOTAL EQUIPMENT COST			R	12,332.60	
1.3	SITE ESTABLISHMENT					
1.3.1	Site office & kitchen, change room & toilets			R	16,800.00	Based on R120/m² for 140m² (wall fence)
1.3.2	Site fencing & security			R	-	Currently available
1.3.3	Site infrastructure			R	-	Currently available
1.3.4	Storage facility - dried product			R	55,165.11	Allowance for 40% product storage based on R40/m²
1.3.5	Storage Land -Raw material			R	-	Currently available
1.3.6	Direct heat Drying facility		65	R	4,875.00	R75/m²
				R	76,840.11	
1.4	RUNNING COSTS					
1.4.1	General labor		19	R	34,710.72	Based on 8 hr shift per day for 22 days @R10.38/h
1.4.2	Site Manager		1	R	7,157.92	Based on 8 hr shift per day for 22 days @R 40.67/h
1.4.3	Assistant Site Manager		1	R	-	Based on 8 hr shift per day for 22 days
1.4.4	Security guard		1	R	-	Currently available
1.4.5	Water, electricity & municipal rates		1	R	250.00	Based on Max 10kL/month
1.4.6	Fuel		1	R	1,800.00	Based on 1277km per month @R1.41/km
1.4.7	Vehicle lease contract(Full maintenance)		1	R	2,380.00	Bakkie & Trailer on a full maintenance lease
1.4.8	Packaging		28732	R	-	No packaging required
1.4.9	Raw material		1	R	627.00	Allowance for 15% handling fee on transport
1.4.10	Drying cost		3539.04	R	-	This cost is not incurred
				R	46,925.64	
2				PRODUCT VALUE		
2.1	Briquette price @10.9MJ/kg for...		40,823	R	92,260.87	Based on R2.26/kg
				R	92,260.87	
3				SUMMARY		
3.1	TOTAL CAPITAL COST	R			95,422.72	Once off
3.2	TOTAL RUNNING COST	R			46,925.64	Monthly
3.3	TOTAL ESTIMATED REVENUE	R			92,260.87	Monthly
3.4	CONTIGENCIES	R			11,450.73	Once off, usually 12% of the total project cost

Brick Area(m²)	Monthly Prod(kg/m)	Brick mass(kg)	
0.010000	49,694		0.118403513
No .bricks	Required drying area(m²)		
344,782	3,448		

Table 7.4: Summary of costing for the Bikernmayer model

The above model, which follows the methods suggested by Peters & Timmerhaus (1999), is based on a production capacity of **596.3 tons per year** over a project life of **five (5) years**. The total fixed capital investment of this project is **R 106, 874** with a running cost of **R 733,317** per annum and annual revenue of **R1, 306,664**. According to Peters & Timmerhaus (1999), the annual rate of return which is greater than 9% is considered attractive for investors. Therefore an annual rate of return will be used as one of the key indicators to evaluate the economic viability of this project.

7.4.2 Depreciation

The total fixed capital of R 106,874 including 12% contingency is depreciated continuously over a period of five years based on equation 7.3.1.1(a) and an annual interest rate of 10.5%. Companies pay tax on their taxable income at a flat rate. That rate is 30% for financial years ending between 1 April 2004 and 31 March 2005, and 29% for 1 April 2005 to date, retrieved February 16, 2010 from the World Wide Web:

http://www.sars.gov.za/Tools/Documents/DocumentDownload.asp?FileID

$$d_n = \left(\frac{1}{(1+i)^n} \right)$$

$$d_n = \left(1 \to 0.9049 \to 0.8190 \to 0.7412 \to 0.6707 \to 0.6071 \right)$$

$$\therefore \overbrace{R347{,}341}^{n=0} \to \overbrace{R0}^{n=1} \to \overbrace{R0}^{n=2} \to \overbrace{R0}^{n=3} \to \overbrace{R0}^{n=4} \to \overbrace{R0}^{n=5}$$

7.4.2 Inflation

An inflation index is calculated as follows:

$$index = (1+i)^n$$
$$for \_ n = 0.....5;$$

$$(index)_0 = (1+0.105)^0 = 1$$
$$(index)_1 = (1+0.105)^1 = 1.105$$
$$(index)_2 = (1+0.105)^2 = 1.221$$
$$(index)_3 = (1+0.105)^3 = 1.349$$
$$(index)_4 = (1+0.105)^4 = 1.491$$
$$(index)_5 = (1+0.105)^5 = 1.647$$

At year zero, the fixed capital and running costs are being used to purchase new equipment construct the plant and procurement of all the necessary resources required for start-up and commissioning. At this stage, the project is not expected to be making any revenue and thus have a negative cash flow. A total capital of R106, 874 and a running cost of R 563,108 is spent in year zero for preparation of the plant, resulting in the total expenditure of R 669,981. A tax break is applied in this instance due to the fact that the project has not made any revenue, a tax break will be applied for as long as the project is yielding a negative cash flow. The project results with a net income of –R669, 981 at the end of year zero.

In year one (1), the project is assumed to be fully completed, with all equipment installed and commissioned. A total revenue of R 1,107,130 is expected by the end of year one. The running cost is inflated by inflation index 1 as follows:

$$Running(cost)_1 = R563,108 \times 1.105 = R622,234$$

There is no fixed capital to be depreciated in year one as it has all been spent in year zero. The gross income before tax includes the negative cash flow from the previous year and is calculated as follows:

$$\text{Net Income}_{(\text{before tax})} = \text{Revenue} - \text{Running}_{(\text{cost})} + \text{cash flow}_{(\text{year}_0)}$$

$$\therefore \text{Net Income}_{(\text{before tax})} = R1,107,130 - R622,234 + (-R669,981) = -R123,570$$

The above calculation indicates a net negative cash flow at the end of year, however the fixed capital investment is written off during the first few months of the second half of year and the overall negative cash flow has mainly been influenced by the running cost. Although the project has passed the break- even point in the second half of year one, the overall net income is negative and not subject to a tax rate of 29% as stipulated in the South African tax law for businesses.

Similarly, the revenue and running costs are inflated in year two. This results in a net positive cash flow of R 615,807 as shown in table 7.5. Since the net cash flow is positive, an income tax rate of 29% is applied on the gross income (-R178, 584) resulting in the net cash flow of R437, 223.

7.4.4 Internal Rate of Return (IRR)

The compounded interest rate that, when used to discount the cash flows, will yield the present value equal to the initial capital investment, it is known as the internal rate of return. Therefore the internal rate of return is simply the interest rate that will make the present value of the cash inflows equals to the present value of the cash outflows. Internal Rate of Return is mathematically defined as follows:

$$f(i) = \frac{C_1}{(1+i)^1} + \frac{C_2}{(1+i)^2} + \frac{C_3}{(1+i)^3} + \frac{C_4}{(1+i)^4} + \frac{C_5}{(1+i)^5} + \frac{C_6}{(1+i)^6} - Investmet$$

$find : i, for : f(i) = 0$

$$\therefore 0 = \frac{-669,981}{(1+i)^1} + \frac{-123,570}{(1+i)^2} + \frac{615,807}{(1+i)^3} + \frac{817,011}{(1+i)^4} + \frac{902,797}{(1+i)^5} + \frac{997,591}{(1+i)^6} - 106,873$$

$Solving\_numerically$

$\therefore i = 31.4\%$

7.4.5 Net present value (NPV)

This is another method that takes into account the time value of money when evaluating investment proposals. It assumes that the rate at which the capital can be borrowed is known, and then discounts all cash flows at this rate back to the present. The net present value of all the cash flows is then compared with the investment necessary to produce the inflows. Concisely, the net present value (NPV) of an investment proposal is the present values of all future cash inflows (PV_{in}) less the investment outlay (I_{out}). Net Present Value method (NPV) is mathematically defined as follows:

$$NPV = PV_{in} - I_{out}$$

$$PV = \frac{C_1}{(1+i)^1} + \frac{C_2}{(1+i)^2} + \frac{C_3}{(1+i)^3} + \frac{C_4}{(1+i)^4} + \frac{C_5}{(1+i)^5} + \frac{C_6}{(1+i)^6} - I_{out}$$

$$\therefore NPV = \frac{-669,981}{(1+i)^1} + \frac{-123,570}{(1+i)^2} + \frac{615,807}{(1+i)^3} + \frac{817,011}{(1+i)^4} + \frac{902,797}{(1+i)^5} + \frac{997,591}{(1+i)^6} - 106,873$$

$$NPV = R786,769 - R106,873 = R679,896$$

7.4.6 Profitability Index (PI)

As discussed under section 7.1.3, equation 7.4 is used to calculate the profitability index as follows:

$$PI = \frac{NPV - I_{out}}{I_{out}}$$

$$\therefore PI = \frac{R676,896 - R106,874}{R106,874} = \underline{\underline{5.33}}$$

7.4.7 Gross profit margin

Gross profit margin is a financial ratio used to assess the profitability of project core activities, excluding fixed costs. It is a measure of how well each rand of a project's revenue is available to meet expenses and profits after paying for the goods or services that were sold.

The general calculation is:

$$GPM = \frac{R \quad C_s}{R} \dots\dots\dots\dots\dots\dots\dots\dots\dots\dots\dots\dots\dots\dots\dots\dots\dots\dots\dots7.8$$

Gross Profit Margin (GPM)

Revenue(R)

Cost of Sales (C_s)

According to equation 7.8, the gross profit margin is calculated as follows:-

$$GPM = \frac{R1,306,664 - R733,317}{R1,306,664} \times 100\%$$

$$\therefore GPM = \underline{\underline{44\%}}$$

The gross profit margin is related to the net profit margin, which assesses the profitability of project after including fixed costs. Gross profit margin indicates the relationship between net sales revenue and the cost of goods sold. A high gross profit margin indicates that a business can make a reasonable profit on sales, as long as it keeps overhead costs in control.

7.4.8 Payback period

The payback method is one of the traditional approaches and is still commonly used, despite the fact that is decidedly inferior to the IRR and NPV methods. The payback method simply gives an indication of how soon the initial investment will be paid back, however it is not based on detailed consideration of the time value of money via compound interest calculations. The payback period (p_{period}) is mathematically defined as the ratio of depreciable fixed capital (K_f) and the average profit per year (Profit) over the useful life of the project.

$$p_{period} = \frac{K_f}{\text{Pr}\, ofit}$$

$$\therefore p_{period} = \frac{R106,874}{R262,171} = \underline{\underline{0.408\, years}}$$

The above listed economic evaluation methods are only shown for the tabulated summary is presented for the Porta press and screw

Bikernmayer model,

SUMMARY CALCULATIONS		Year 0	Year 1	Year 2	Year 3	Year 4	Year 5
Project Life(years)		0	1	2	3	4	5
Inflation index		1.00	1.11	1.22	1.35	1.49	1.65
Normal income Tax for businesses	29.00%						
Fixed Interest rate(Prime rate)	10.50%						
Revenue		R 0	R 1,107,130	R 1,351,834	R 1,493,777	R 1,650,623	R 1,823,939
Carbon Credits @ R206.43/Ton CO₂		R 0	R 61,515	R 75,111	R 82,998	R 91,713	R 101,342
Total Revenue	R 1,306,664	R 0	R 1,168,645	R 1,426,945	R 1,576,774	R 1,742,336	R 1,925,281
Fixed Capital		106,873.44	R 0	R 0	R 0	R 0	R 0
Running Cost(R/year)	R 733,317	R 563,108	R 622,234	R 687,569	R 759,763	R 839,538	R 927,690
Depreciated capital(R/year)		R 106,873	R 0	R 0	R 0	R 0	R 0
NET INCOME BEFORE TAX							
Taxable income	R 423,276	R -669,981	R -123,570	R 615,807	R 817,011	R 902,797	R 997,591
Tax @ 29%		R -	R -	R -178,584	R -236,933	R -261,811	R -289,301
NET INCOME AFTER TAX	262,171	R -669,981	R -123,570	R 437,223	R 580,078	R 640,986	R 708,290
PV of Net Cash Flow	783,769	R -669,981	R -111,828	R 358,079	R 429,932	R 429,932	R 389,078
Iteration target set to zero(0)	0.00						
Gross Margin(%)	44%						
Profitability Index (PI)	3.39						
Internal Rate of Return IRR(%)	31.4%						
NPV (ZAR)	R 676,896						
Return Period(yrs)	0.40%						

Table 7.9: Summary of calculation for economic evaluation for the Bikernmayer conceptual model

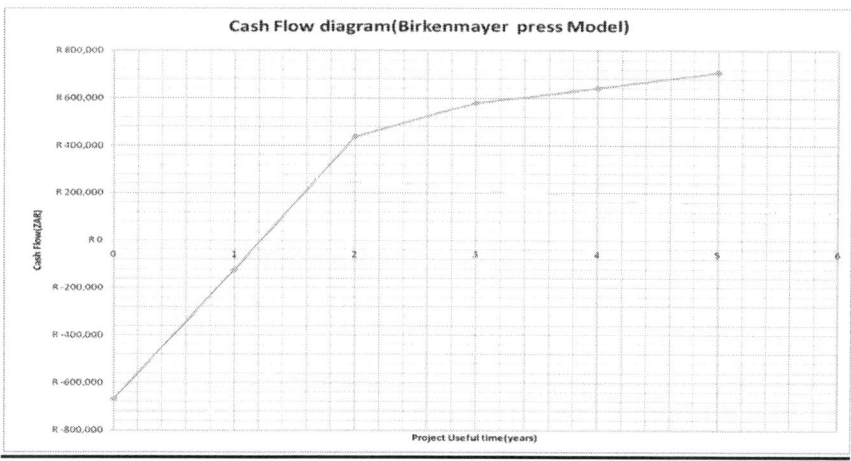

Table 7.2: The cash flow diagram for the Bikernmayer press model

The above diagram shows the relationship between the nett income and the project period, it is evident the project breaks even just after year 1. The project is expected to generate a positive income after the breakeven point.

7.4.9 Carbon Credits

There are two distinct types of Carbon Credits: Carbon Offset Credits and Carbon Reduction Credits. Carbon Offset Credits consist of clean forms of energy production, wind, solar, hydro and biofuels. Carbon Reduction Credits consists of the collection and storage of Carbon from our atmosphere through reforestation, forestation, and ocean and soil collection. Both approaches are recognized as effective ways to reduce the Global Carbon Emissions. Carbon markets and emission trading regimes are part of the Kyoto Treaty, and international carbon emission. Each nation gets a limited amount of carbon credits, which are then made available to individual companies; the companies are then able to sell or buy credits from each other. According to Point Carbon, a website monitoring CO_2 emission trades, an average spot price for a ton-credit of carbon is €19.42 (R206.43) compared to the €40(R425.20) penalty per ton companies face for exceeding limits). The carbon credits increased the overall revenue by 6%.

7.4.9.1 Credits versus Taxes

Carbon credits and carbon taxes each have their advantages and disadvantages. Credits were chosen by the signatories to the Kyoto Protocol as an alternative to Carbon taxes. A criticism of tax-raising schemes is that they are frequently not hypothecated, and so some or all of the taxation raised by a government may be applied inefficiently or not used to benefit the environment. By treating emissions as a market commodity it becomes easier for business to understand and manage their activities, while economists and traders can attempt to predict future pricing using well-understood market theories.

146

7.4.9.2 Advantages of a tradable carbon credit over a carbon tax

The price is more likely to be perceived as fair by those paying it. Investors in credits have more control over their own costs.

- The flexible mechanisms of the Kyoto Protocol ensure that all investment goes into genuine sustainable carbon reduction schemes, through its internationally-agreed validation process.
- If correctly implemented a target level of emission reductions is achieved with certainty, while under a tax the actual emissions would vary over time.
- It provides a framework for rewarding people or companies who plant trees or otherwise sequester carbon.

7.4.9.3 The advantages of a carbon tax

- Possibly less complex, expensive, and time-consuming to implement. This advantage is especially great when applied to markets like gasoline or home heating oil.
- Perhaps some reduced risk of certain types of cheating, though under both credits and taxes, emissions must be verified.
- Reduced incentives for companies to delay efficiency improvements prior to the establishment of the baseline if credits are distributed in proportion to past emissions.
- When credits are grandfathered, this puts new or growing companies at a disadvantage relative to more established companies.

- It is clear what effect the policy has on the price of energy.

7.4.9.4 Carbon Credits Calculations

Given mass of the fuel $m_f = 524g$ and mass of carbon dioxide produced $m_{CO_2} = 293.32g$

Carbon dioxide loading factor is obtained as:

$$\therefore \hbar = \frac{m_{CO_2}}{m_f} = \left(\frac{293.32g}{524g}\right) = \underline{0.5598}$$

Total mass production as per proposed in the Bikernmayer model is <u>35,490 kg per month</u>. The estimated quantity of carbon dioxide produced from briquettes is calculates as follows.

$$m_{CO_2} = p_r \times \hbar$$
$$m_{CO_2} = 44,362kg.month^{-1} \times 0.5598$$
$$\therefore m_{CO_2} = 24,33.8kg / month$$
$$C_{cr} = R206.43.ton^{-1} \times 24.834ton.month^{-1} = (R5,126.45 / month) \rightarrow \underline{\left(R61,514.7 / year\right)}$$

7.4.9.5 Product Demand

The proposed site is located in Tsakani with over 200,000 population size; the site has average area of $4,655m^2$ with 80% of the total area available for development. There are over 432 homes within 80m radius from the site. Figure 7.4.8.5(a) shows a satellite landscape image of the proposed site (Thuthukani) with the highlighted area within 80 meter radius from the production facility, the selected area is practical for manual transportation of products using wheel barrows. Based on the market survey conducted (March 2009), each home uses 2kg of wood per day to start coal stoves or other combustion appliances in summer. An average of 4kg of fuel wood is used by each home during the winter season. Two (2) kilograms of fuel wood is equivalent to 13 eco- fuel briquettes, resulting in the total demand of 168,480 briquettes per month in summer and twice the demand (336,960) in winter. The project is aimed at producing 28,620 briquettes per month, which is only sufficient to cater for 17% of the consumers within the defined 80m radius from the production site. This reduces the risk of overproduction and the cost of final product storage. The production capacity can be expanded once a positive response relating to consumer demand has been identified.

Figure 7.4.9.5(a): Google earth satellite image for the proposed site in Tsakane, $26^0 20^{'} 56.03^{''} S$, $28^0 22^{'} 28.36^{'} E$, altitude=1637m.

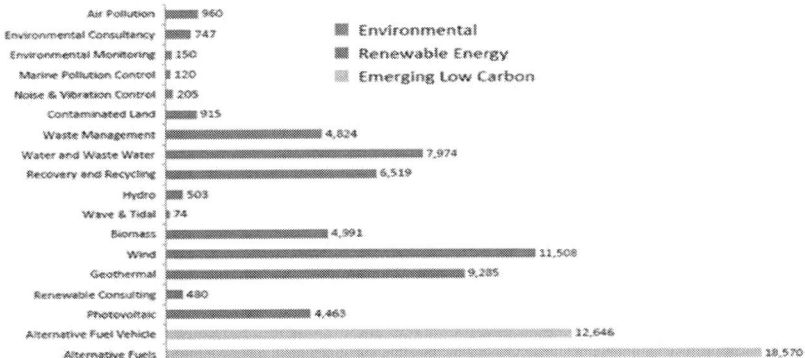

UK market value of Level 2 products and services, 2007/8 (£m)

Figure 7.10: United Kingdom market value of level 2 products and services based on 1£=R12.07

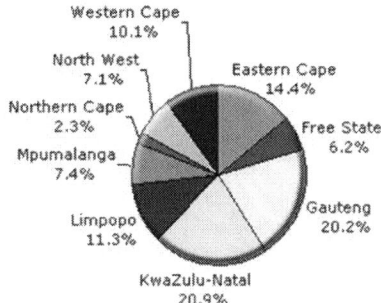

Population by province

Western Cape 10.1%
North West 7.1%
Northern Cape 2.3%
Mpumalanga 7.4%
Limpopo 11.3%
KwaZulu-Natal 20.9%
Eastern Cape 14.4%
Free State 6.2%
Gauteng 20.2%

Figure 7.11: South African population distribution per province

7.5. The cost of energy from various sources

The electricity cost of 45 cents per kilowatt hour was estimated based on the market survey findings conducted in March 2009. The figures presented in table 7.3 indicate that coal is presently the cheapest energy source followed by electricity at 10.7cents per Mega joule and 12.5 cents per Mega joule respectively. The price of coal in the local market is usually regulated by the local distributors at an average rate of 13% per annum; furthermore ESKOM has confirmed a 34% increase in electricity price, resulting in the most reliable energy sources becoming more expensive. The eco-fuel briquettes is the source of renewable energy source and cost R1.05/kg to produce and yet the cheapest renewable energy source.

Type	Electricity	Coal	Wood	Paraffin	LP Gas	Cobble stones	Eco-fuel briquettes
Rate	R0.45/kW.h	R2.60/kg	R2.80/kg	R13.50/kg	R15/kg	R30.55/kg	R2.26/kg
Energy content	3.6MJ/kWh	24.3MJ/kg	16MJ/kg	42MJ/kg	49.3MJ/kg	19.7MJ/kg	18.9MJ/kg
Fuel cost	12.5Cents/MJ	10.7Cents/MJ	17.5 Cents/MJ	32 Cents/MJ	30.4 Cents/MJ	155Cents/MJ	11.9Cents/MJ

Table 7.3: Cost of energy from various sources

7.6 Conclusions

Following market sizes based on the United Kingdom in figure 7.10 and the South African market being quantified as one tenth of the United Kingdom for comparison, the potential biomass, renewable and alternative fuels market in South Africa at 10 percent of £28.7 billion. This is equivalent to R34.7 billion potential market in South Africa, it is therefore fairly clear that the waste to energy market is considerable. The waste to energy sector in South Africa is at the earliest stage and therefore rapid growth should be experienced in the decade to come.

Based on the populations data obtained from statistics South Africa, the average population of the country is currently 48.6 million with Gauteng province contributing 20.2 %(9.9 million). An estimated potential market in Gauteng province is about R7 billion, and the population size of Tsakane Township is approximately 200,000 (2% of Gauteng province), resulting in a potential market of R141 million. The Thuthukani project will service 432 homes, and with an average 7 persons per home, the market is about R2.11 million per annum. This can be compared to the expected sale of briquettes which is R0.43 million per annum, which is in the right order of magnitude.

Principal feedstock are biomass such agricultural waste, coffee grounds, saw dust, cow dung, waste paper and alien vegetation with a calorific value of about 14MJ/kg, and domestic coal fines with a calorific value of about 21.33 MJ/kg. Combination of the above based on each defined recipe dynamics to have an average caloric value of 18-20MJ/kg.

A feed-in tariff or renewable energy payments is a policy mechanism designed to encourage the adoption of renewable energy sources. It typically includes three key provisions: guaranteed grid access, long-term contracts for the electricity produced, and purchase prices that are methodologically based on the cost of renewable energy generation. Under a feed-in tariff, an obligation is imposed on regional or national electricity utilities to buy renewable electricity (electricity generated from renewable sources, such as solar thermal power, wind power, wave and tidal power, biomass, hydropower and geothermal power), from all eligible participants. The renewable energy feed in tariff (REFIT) is being concluded by the National Energy Regulator of South Africa (NERSA) in October 2009 at R1.18 kW.hr (Chikanga, 2009). This simply means that one kilogram of eco-fuel briquettes with calorific value of $18.9 MJ.kg^{-1}$ can be sold at R6.20 which is three times the current cost of ESKOM electricity.

The current economic model is based at a selling price R2.26 per kilogram for the ease of accommodating the local market. This represents a lower cost per MJ than electricity. The financial model for both Porta press and screw press are not economically viable as running cost is greater than the gross project revenue.
The Capital investment details as follows:

- Operating life time: 5years
- Direct Capital is R 106,873.44 + VAT
- Direct operating costs (also to be capitalised) at R563,108 + VAT,
- Total Capital investment is at R669,981+ VAT
- Gross Profit margin : 44%
- Profitability index of: 5.33
- Internal Rate of return :31.4%

- Net Present value: R676,896

- Return period: 0.408 years

The customer profile as currently at hand is 17% of the selected area within 80m radius from production site. The remaining 83% will be in need of energy as they become aware of the new product offering. The principal driver for this project is socio economic development and it is being strengthened by Eskom's inability to provide sufficient energy. As secondary driver is the global drive to reduce emissions and fossil fuel usage, this technology does exactly this whilst diverting waste from landfill. In the Polokwane declaration (2008), it is stated that South Africa will have no calorific waste to landfill by 2014. (Van der Merve, 2009), hence legislation will also provide a major part of the drive.

7.6 References

Allen, D. 1991. *Economic Evaluation of Projects*, 3rd Edition, IChemE.

Coulson, J.M., Richardson, J.F. & Sinnott, R.K. 1999. *Chemical Engineering - Volume 6: An Introduction to Chemical Engineering Design 3rd ed*, Butterworth Heinemann.

Chikanga, D. 2009. Waste-to-energy could alleviate pressure on SA landfills. *Engineering News* (October 18).

Garrett, D.E. 1989. *Chemical Engineering Economics*, Van Nostrand Reinhold.

Google earth World Wide Web.23 February 2010. http://earth.google.com/

IChemE, 2000. *A Guide to Capital Cost Estimating*, 4th Edition, IChemE.

Perry, R.H. & Green, D. 1998. *Perry's Chemical Engineer's Handbook*, 7th Edition, McGraw-Hill., section 9.

Peters, M.S. & Timmerhaus, K.D. 1991. *Plant Design and Economics for Chemical Engineers*, 4th Edition, McGraw-Hill, chapters 6-9.

South African revenue service. 16 February 2010.
http://www.sars.gov.za/Tools/Documents/DocumentDownload.asp?FileID

Ulrich, G.D. 1984. *A Guide to Chemical Engineering Process Design and Economics*, Wiley, chapters 5-8.

Van der Merve, C. 2009. NERSA makes Refit phase two decision; PPA expected in November, Engineering News (November 30).

7.7. Nomenclature and Units

Symbol	Definition	SI units
P_v	Present Value	ZAR
f_v	Future value	ZAR
i	Fraction of annual interest rate	–
PI	Profitability index	–
l_{out}	Capital Investment	ZAR
d_n	Fraction of discount rate	–
IRR	Internal rate of return	%
NPV	Net present value	ZAR
p_{period}	Payback period	years
$C_{1..n}$	Cash flow	ZAR
K_f	Depreciated fixed Capital	ZAR
GPM	Gross profit margin	%
R	Revenue	ZAR
C_s	Cost of sales	ZAR
COC	Carbon offsets credits	ZAR
CRC	Carbon reduction credits	ZAR
v_s	Salvage value	ZAR

n	Service value	*years*
REFIT	Renewable energy feed in tariff	*ZAR*

Nomenclature and units (Continued)

m_f	Mass of fuel	*kg*
m_{CO_2}	Mass of carbon dioxide	*kg*
p_r	Production rate	*kg / month*
$\hbar$	Carbon dioxide loading factor	–
C_{cr}	Carbon credits	*ZAR*

n_b	**Number of briquettes**	–
d_r	Daily production rate	*kg / day*
m_r	Monthly production rate	*kg / month*
a_r	Annual production rate	*tons / year*
$m_{r(dr)}$	Monthly production rate(dried product)	*dry (kg) / month*

CHAPTER 8: Environmental impact

8.1 Background

The potential health hazard to an individual by a material used in any chemical process is a function of the inherent toxicity of the material and the frequency and duration of exposure. It is common practice to distinguish between the short-term and long-term effects of a material. A highly toxic material that causes immediate injury is classified as a safety hazard while a material whose effect is only apparent after long exposure at low concentrations is considered as an industrial health and hygiene hazard (Lewis, 1997).

The permissible limits and the precautions to be taken to ensure that such limits will not be exceeded are quite different for these two classes of toxic materials. The inherent toxicity of a material is measured by tests on animals. The short-term effect is expressed as LD_{50}, the lethal dose at which 50 percent of the test animals do not survive. Estimates of the LD_{50} value for humans are extrapolated from the animal tests. On the other hand, the permissible limits of concentration for the long-term exposure of humans to toxic materials are set by the threshold limit value (TLV). The latter is defined as the upper permissible concentration limit of the material believed to be safe for humans even with an exposure of 8 hr per day, 5 days per week over a period of many years (Wright & Welbourn, 2002).

Recommended TLV values are published in the hazardous substance database bank by the Occupational Safety and Health Agency (OSHA), for simplicity proposes the OSHA data base will be assumed appropriate for the South African context. With the uncertainties involved in the designation of occupational exposure standards and the variability of the occupational environment, it would be unreasonable to interpret occupational limits as rigidly as one might interpret an engineering standard or specification. Fortunately, there has been a recent effort to make these rather subjective judgments more scientific and uniform by the application of statistics. The latter makes it possible to develop decision-making strategies that can prescribe how many samples to take, where and when to take them in the workplace, and how to interpret the results (OSHA, 1990).

CARBON MONOXIDE CASRN: 630-08-0 **OSHA Standards:** Permissible Exposure Limit: 8-hr Time Weighted Avg: 50 ppm (55 mg/cu m). [29 CFR 1910.1000 (7/1/98)] Vacated 1989 OSHA PEL TWA 35 ppm (40 mg/cu m); Ceiling limit 200 ppm (229 mg/cu m) is still enforced in some states.
NITRIC OXIDE CASRN: 10102-43-9 **OSHA Standards:** Permissible Exposure Limit: 8-hr Time Weighted Avg: 25 ppm (30 mg/cu m). [29 CFR 1910.1000 (7/1/98)]
HYDROGEN SULFIDE CASRN: 7783-06-4 **OSHA Standards:** Permissible Exposure Limit: Acceptable Ceiling Concentration: 20 ppm. [29 CFR 1910.1000 (7/1/98)] Permissible Exposure Limit: Acceptable maximum peak above the acceptable ceiling concentration for an 8-hour shift. Concentration: 50 ppm. Maximum Duration: 10 minutes once, only if no other meas. exp. occurs.
SULFUR DIOXIDE CASRN: 7446-09-5 **OSHA Standards:** Permissible Exposure Limit: 8-hr Time Weighted Avg: 5 ppm (13 mg/cu m). [29 CFR 1910.1000 (7/1/98)]
CARBON DIOXIDE CASRN: 124-38-9 **OSHA Standards:** Permissible Exposure Limit: 8-hr Time Weighted Avg: 5000 ppm (9000 mg/cu m). [29 CFR 1910.1000 (7/1/98)]
NITROGEN DIOXIDE CASRN: 10102-44-0 **OSHA Standards:** Permissible Exposure Limit: Ceiling value: 5 ppm (9 mg/cu m). [29 CFR 1910.1000 (7/1/98)]

Table 8.1: Occupational Exposure standard (source: OSHA, 1990)

CO$_2$(ppm)	CO(ppm)	H$_2$S (ppm)	SO$_2$ (ppm)	NO (ppm)	NO$_2$ (ppm)	
5000 ppm	50 -200ppm	20 ppm.	5 ppm	25 ppm	5 ppm	C_{lim}
213,328.78	73.78	4.32	3.67	1.34	2.73	C_x

Table 8.2: Actual emissions gas concentration for the selected toxic gases vs. the maximum exposure limit over 8 hrs.

Table 8.2 shows a comparison between the actual gas emissions of eco-fuel briquettes and maximum human exposure limit over 8 hours. The above table clearly indicates that the gas emission produced by the eco-fuel briquettes conforms by 83.3 % to the Occupational Safety and Health Agency (OSHA) occupational exposure standards. The carbon dioxide concentration is 46 times above the human exposure limit over the 8 hour exposure time. The safe exposure time for this high carbon dioxide concentration was calculated as follows:

$$t_{exp} = \frac{C_{lim}(ppm)}{C_x(ppm)} \times t_{lim}$$

$$\therefore t_{exp} = \frac{5,000\,ppm}{213,328.78\,ppm} \times 8hr = 0.188hr = \underline{\underline{(11.25\,min)}}$$

Although the other gases are in low concentrations, the maximum safe exposure time to the combustion gasses produced by the eco-fuel briquettes is 11.25 minutes due to the high carbon dioxide concentration. In principle, the fuel briquette is deemed a clean fuel compared to other fossil fuels. The carbon dioxide produced is 2892 times the carbon monoxide, which indicates a complete combustion. Furthermore, Carbon dioxide is less toxic than carbon monoxide; therefore combustion of eco-fuel briquettes is even more suitable for indoors well-ventilated applications.

Flue Gas Composition

Fuel Choices & OSHA: Chemical Species	OSHA TWA *ceiling, ppm	Natural Gas	Fuel Oil	Coal
Nitrogen, N_2		78-80%	78-80%	78-80%
Carbon dioxide, CO_2	5000	10 – 12%	12-14%	
Oxygen, O_2		2-3%	2-6%	7%
Carbon monoxide (CO)	50	70-110ppm	70-160ppm	
Nitrogen oxides (NO_x)	NO-25, NO_2-5*	50-70ppm	50-110ppm	1%
Ammonia, NH_3	50	Used in removal of NO_x.		
Sulphur dioxide (SO_2)	H_2S-20*, SO_2-5		180-250ppm	>2,000ppm
Hydrocarbons (C_xH_y)			<60ppm	
Mercury, Hg				>200lb/year/plant
Fly Ash		none	minimal	12%

Table. Summary of flue gas composition ranges for power plants fueled by gas, oil and coal. Given these inconvenient contaminants it is no surprise that EOR by flue gas injection has been discontinued, sometimes converted to nitrogen injection, in most projects which attempted that EOR implementation. OSHA's TWA limits are allowed for 8-hour personnel shifts. OSHA's Ceiling limits should not be exceeded at any time for personnel.

Table 8.3: Source: Jim Meyers MPE; Flue gas green house and EOR, 2009-09-07

Table 8.4 summarises the flue gas compositions for Natural gas, fuel Oil and coal; the above results indicate that the eco-fuel briquettes flue gas is cleaner than the above listed fossil fuels.

Chemical species	OSHA-max(limit)- (ppm)	Natural gas	Fuel oil	Coal	Eco-fuel briquettes
Nitrogen (N_2)	-	78-80%	78-80%	78-80%	60.3%
Carbon dioxide(CO_2)	5,000ppm	10-12%	12-14%	10.6%	21.3%
Oxygen (O_2)	-	2-3%	2-6%	7%	12.8%
Carbon Monoxide (CO)	50-200ppm	70-110ppm		5,579ppm	74ppm
Nitrogen dioxide (NO_2)	5ppm			1%	2.73ppm
Nitric Oxide (NO)	25ppm			1%	1.34ppm
Ammonia (NH_3)	50ppm				
Sulphur dioxide (SO_2)	5ppm			>2,000ppm	3.67ppm
Hydrocarbon (C_xH_y)	-				
Hydrogen Sulphide (H_2S)	20ppm				4.32ppm
Mercury (Hg)	-	-			
Ash	-	0	0	12%	10.46%

Table 8.4: Comparison of flue gas quality of various fuels

8.2 The fuel briquette and the Environment

Ultimate analysis: Product						
	Carbon	Hydrogen	Oxygen	Nitrogen	Sulphur	Calorific Value
Eco-fuel briquette	36.65%	4.60%	36.30%	0.75%	0.34%	18.9 MJ/kg

Proximate Analysis: Product					
	Ash	Moisture	Fixed Carbon	Volatile Matter	Density
Eco-fuel briquette	10.46%	10.90%	26.30%	39.34%	721kg/m^3

Table 8.5: Fuel briquette elemental composition measured using ICP-Optima model 2100DV

The above test results indicate that the eco-fuel briquette has less that 1.5 % of nitrogen and sulphur which are the common elements that heavily contribute to environmental pollution when combusted in the presence of atmospheric oxygen.

8.2.1 Experimental air requirement

A theoretical air requirement for combustion for the eco-fuel briquettes was calculated by assuming ideal gas behaviour. The experimental dry air consumed during combustion was measured as 752.65g at absolute pressure of 82.96kPa.

$$PV = nRT$$

$$\therefore V = \frac{nRT}{P} = \frac{\left(\dfrac{752.65g}{28.84g.mol^{-1}}\right) \times 8.314 Pa.m^3.mol^{-1}.K^{-1} \times 413.45K}{82.96 \times 10^3 Pa} = 1.08m^3 \text{ for } 423.3g \text{ fuel burnt}$$

$$\therefore A_g = \frac{1.08m^3}{423.03 \times 10^{-3} kg} = 2.57m^3 / kg$$

8.2.2 Experimental flue gas production

The average volumetric flow rate of the flue gas was measured as $1.08 \times 10^{-4} m^3.s^{-1}$ over a period of 210 minutes and absolute pressure of 82.96kPa.

$$\therefore V = 1.08 \times 10^{-4} m^3.s^{-1} \times 210 \times 60s = 1.36m^3 \text{ for } 423.3g \text{ fuel burnt}$$

$$\therefore f_g = \frac{1.36m^3}{423.03 \times 10^{-3} kg} = 3.20m^3 / kg$$

8.2.3 Description of individual flue gases

8.2.3.1 Nitrogen Oxides (NO_x)

Nitrogen oxides occur in all fossil fuel combustion, through oxidation of atmospheric nitrogen and also from organic nitrogen fuel content and flue gas concentration are enhance by high combustion chamber temperatures. Under normal conditions nitrogen and oxygen do not react with each other. At very high temperatures $800\text{-}1000\,^{\circ}C$ a small proportion of oxygen reacts with nitrogen to give nitrogen oxides. These oxides react with moisture to form a weak solution of nitric acid (Roach, 1992).

Nitric oxide oxidises with time and forms nitrogen dioxide, a brown, toxic, water-soluble gas that can seriously damage lungs, contributes to acid rain and helps to form ozone. With or without Selective Catalytic Reduction (SCR), ammonia ions react with both species:

$$4NH_3 + 6NO \rightarrow 5N_2 + 6H_2O \dots \dots 8.1$$

$$8NH_3 + 6NO \rightarrow 7N_2 + 12H_2O \dots \dots 8.2$$

Use of ammonia in (NO_x) reduction techniques or for flue gas conditioning can have a substantial balance-of-plant impact on coal-fired plants. Ammonia adsorbs on fly ash within the flue as processing system has both the ammonia and ammonium sulphate. This ammonia can desorb during subsequent transport, disposal or use of fly ash. This desorption of ammonia presents several technical and environmental concerns as fly ash occurs in surface of water and landfills. SCR can optimize the NH₃-NO reduction with a minimum of downstream problems developed by ammonia slip. The use of ammonia in Nitrogen oxides reduction may be considered for process that involves combustion of large qualities of fuel briquettes (CCPS, 1989).

8.2.3.2 Carbon Monoxide (CO)

Carbon monoxide is a colourless, odourless gas that is tasteless and non-irritating. It is somewhat less dense than air with a molecular weight of 28 g/mol and, although it is a product of incomplete combustion it is non-flammable. Carbon monoxide like oxygen, has affinity for iron containing molecules, and is about 210 times more effective than oxygen in binding to iron rich haemoglobin in human blood; thus arises its tragic toxicity so often demonstration of suicide (Roach, 1992).

8.2.3.3 Sulphur Oxides $(SO_2 \ \& \ H_2S)$

Almost all hydrogen sulphide oxidises within a day to sulphur dioxide. SO_2 is smelly, toxic, and contributes to acid rain. SO_2 can be removed from flue gas by dry alkaline adsorption before particulate removal. Addition of sodium bicarbonate into flue gas causes it to react in the following manner:

$$2NaHCQ \rightarrow Na_2CO_3 + H_2O + CO_2 \dots\dots\dots\dots\dots\dots\dots\dots\dots\dots\dots\dots\dots\dots\dots\dots\dots\dots 8.3$$

This allows for the sodium carbonate to react with the oxygen and sulphur dioxide in the flue gas to form sodium sulphate and carbon dioxide as follows:

$$Na_2CO_3 + SO_2 + 0.5CO_2 \rightarrow NaSO_4 + CO_2 \dots\dots\dots\dots\dots\dots\dots\dots\dots\dots\dots\dots\dots\dots 8.4$$

With the creation of solid sodium sulphate, the desulfurization of the gas is complete, awaiting capture of solid sodium sulphate particles. In wet limestone scrubbing after particulate removal, limestone slurry in water comes into contact with the flue gas

$$SO_2 + CaCO_3 + H_2O \rightarrow CaSO_3 + H_2O + CO_2 \dots\dots\dots\dots\dots\dots\dots\dots\dots\dots\dots\dots 8.5$$

This calcium sulphite $(CaSO_3)$ is then oxidised to form calcium sulphate, $(CaSO_4)$, gypsum. Contaminants in sheet rock made from recycled gypsum are suspect household environmental hazards

8.3 Discussions

Eco-fuel briquettes are the cleanest solid fuels as compared to other fossil fuels. Composed primarily of 36.65% Carbon, 36.30% Oxygen, 4.6 % Hydrogen, 0.75% Nitrogen, and 0.34% Sulphur and 10.46% Ash as indicated in table 8.5. The main products of the combustion of briquettes are nitrogen, oxygen, carbon dioxide and water vapour, the same compounds we exhale when we breathe. Wood, Coal and Oil are composed of much more complex molecules, with a higher carbon ratio and higher nitrogen and sulphur contents. This means that when combusted, wood, coal and oil release higher levels of harmful emissions, including a higher ratio of carbon emissions, nitrogen oxides, and sulphur dioxide. Wood, Coal and fuel Oil also release ash particles into the environment, substances that do not burn but instead are carried into the atmosphere and contribute to pollution. The combustion of fuel briquette, on the other hand, releases very small amounts of sulphur dioxide and nitrogen oxides, and high ash and carbon dioxide, and very lower levels of carbon monoxide, and other reactive hydrocarbons.

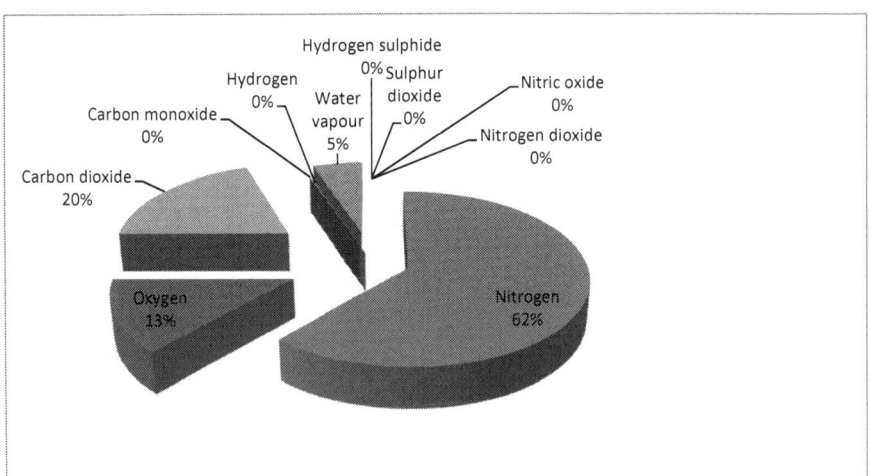

Figure 8.2: Flue gas quality of the fuel briquette measured using gas analyser-Model: Testo 350s

The use of fossil fuels for energy contributes to a number of environmental problems. As the cleaner solid fuels, eco-fuel briquettes can be used in many ways to help reduce the emissions of pollutants into the atmosphere as well as reduction biomass waste disposal to landfills. Burning eco-fuel briquettes instead of other fossil fuels emits fewer harmful pollutants into the atmosphere, and an increased reliance on eco-fuel briquettes can potentially reduce the emission of many of these most harmful pollutants. Pollutants emitted in South Africa, particularly from the combustion of fossil fuels, have led to the development of many pressing environmental problems. Eco-fuel briquettes, emitting fewer harmful chemicals into the atmosphere than other fossil fuels, can help to mitigate some of these environmental issues. These issues include:

- Greenhouse Gas Emissions
- Smog
- Air Quality
- Acid Rain

8.3.1 Greenhouse Gas Emissions

Global Warming or the Greenhouse Effect is an environmental issue that deals with the potential for global climate change due to increased levels of atmospheric greenhouse gases. There are certain gases in the atmosphere that serve to regulate the amount of heat that is kept close to the Earth's surface. Research has indicated that an increase in these greenhouse gases will translate into increased temperatures around the globe, which would result in many disastrous environmental effects. (Fleagle and Businger, 1980).

The principal greenhouse gases include water vapour, carbon dioxide, methane, nitrogen oxides, and some engineered chemicals such as chlorofluorocarbons. While most of these gases occur in the atmosphere naturally, levels have been increasing due to the widespread burning of fossil fuels by growing human populations. The reduction of greenhouse gas emissions has become a primary focus of environmental programs in countries around the world.

One of the principal greenhouse gases is carbon dioxide. Although carbon dioxide does not trap heat as effectively as other greenhouse gases (making it a less potent greenhouse gas), the sheer volume of carbon dioxide emissions into the atmosphere is very high, particularly from the burning of fossil fuels. Because carbon dioxide makes up such a high proportion of the greenhouse gas emissions, reducing carbon dioxide emissions can play a huge role in combating the greenhouse effect and global warming. The combustion of eco-fuel briquettes emits almost twice the carbon dioxide produced by other fossil fuels.

8.3.2 Smog

Smog and poor air quality is a pressing environmental problem, particularly for large metropolitan cities. Smog, the primary constituent of which is ground level ozone, is formed by a chemical reaction of carbon monoxide, nitrogen oxides, volatile organic compounds, and heat from sunlight. As well as creating that familiar smoggy haze commonly found surrounding large cities, particularly in the summer time, smog and ground level ozone can contribute to respiratory problems ranging from temporary discomfort to long-lasting, permanent lung damage. Pollutants contributing to smog come from a variety of sources, including: vehicle emissions, smokestack emissions, paints, and solvents. Because the reaction to create smog requires heat, smog problems are the worst in the summertime.

The use of eco-fuel briquettes does not contribute significantly to smog formation, as it emits low levels of nitrogen oxides, and virtually no particulate matter(refer to table 8.2 & 8.4). For this reason, it can be used to help combat smog formation in those areas where ground level air quality is poor. The main sources of nitrogen oxides are burning of coal, motor vehicles, and industrial plants. Increased eco-fuel briquettes use in the townships and informal settlements, could serve to combat smog production, especially in semi-urban centres where it is needed the most. Particularly in winter, when energy demand high and smog problems are the greatest, communities and schools could use eco-fuel briquettes to fuel their operations instead of other more polluting fossil fuels. This would effectively reduce the emissions of smog-causing chemicals, and result in clearer, healthier air around townships.

8.3.3 Air quality

Particulate emissions also cause the degradation of air quality in the Gauteng. These particulates can include soot, ash, metals, and other airborne particles. A study by the union of concerned scientists in 1998, entitled Cars and Trucks and Air Pollution, showed that the risk of premature death for residents in areas with high airborne particulate matter was 26 percent greater than for those in areas with low particulate levels. Eco-fuel briquettes may potentially emits a significant amount of particulates into the atmosphere: The eco-fuel briquette contains 10.46% of ash, when burning eco-fuel briquettes, more attention should be paid in terms of managing the potential particulate matter.

8.3.4 Acid Rain

Acid rain is another environmental problem that affects most of the areas in South Africa, damaging crops, forests, wildlife populations, and causing respiratory and other illnesses in humans. Acid rain is formed when sulphur dioxide and nitrogen oxides react with water vapour and other chemicals in the presence of sunlight to form various acidic compounds in the air. The principal source of acid rain causing pollutants, sulphur dioxide and nitrogen oxides, are coal fired power plants. Since eco-fuel briquettes emit virtually no sulphur dioxide, and up to 80 percent less nitrogen oxides than the combustion of coal, increased use of eco-fuel briquettes could provide for fewer acid rain causing emissions.

8.4 Conclusions

The production of eco-fuel briquettes is primarily targeting the local market with the aim of creating a basic platform for socio-economic development by creating jobs and providing necessary skills to the communities. The project has environmental benefits such as reducing the amount of general biomass waste going into the municipal landfills. As the product is considered a renewable energy source, the project has also carbon offset credits which may be earned in the form of revenue or tax credits. The health and safety of consumers is not compromised as the briquettes burn cleanly within the acceptable emission exposure limits set out by OSHA. However, consumers need to be educated of the health hazards associated with the exposure of combustion gasses over prolonged periods without sufficient ventilation and the dangers related with the use of uncertified combustion appliances.

Most of heating appliances rely on the availability of air within the living space for the supply of combustion air. This includes unvented heaters, coal stoves, and other combustion appliances that also vent combustion products into the indoors. However, if the home is insulated and the air tightness is increased, then these appliances are polluting the home and should be vented to the outside. Combustion devices should not only be fitted with a direct exhaust (chimney), but also with a dedicated intake, in order to supply the unit with combustion air. Ventilation is a necessary requirement for the proper operation of any home. Many existing homes rely on natural ventilation to control interior humidity, odours and combustion products. Natural ventilation is wind dependant and therefore unreliable. The quality of construction, however, is improving, and homes are becoming tighter and more energy efficient. These houses cannot rely on natural ventilation and as a result new residential buildings are developing a need for mechanical ventilation.

8.5 References

CCPS (1989) *Guidelines for Chemical Process Quantitative Risk Analysis*, Center for Chemical Process Safety (CCPS) of the AIChE.

Fleagle, R.G. and Businger, J.A., 1980.*An introduction to atmospheric physics*, 2nd edition.

Lewis, R.J. 1997. *Sax's Dangerous Properties of Industrial Materials, 6th ed*, Van Nostrand Reinhold.

LeVine, R., 1988. *Guidelines for Safe Storage and Handling of High Toxic Hazard Materials*, Center for Process Safety of the AIChE.

Roach, S.A., 1992. *Health Risks from Hazardous Substances at Work: Assessment, Evaluation and Control*, Pergamon.

Reis A., Smith, I., Peube, J. L., Stephen, K., Ed. Energy Economics and Management in Industry, Proceedings of the European Congress, Portugal ,2-5 April 1984, Vol. 1, *Energy Economics*, p. 103.

Roach, S.A., 1992. *Health Risks from Hazardous Substances at Work: Assessment, Evaluation and Control*, Pergamon.

OSHA, 1990. *OSHA Regulated Hazardous Substances: Health, Toxicity, Economic and Technological Data*, U.S. Dept. of Labour.

Wright DA, & Welbourn P, 2002. *Environmental toxicology*, Canada: Cambridge University press.

8.6. Nomenclature and Units

Symbol	Definition	SI units
R	Ideal gas constant	$m^3 .Pa .mol^{-1} .K^{-1}$
t_{exp}	Actual exposure time	s
t_{lim}	Exposure time limit	s
A_g	Theoretical combustion air requirements per kg of fuel	m^3 / kg_{fuel}
f_g	Theoretical flue gas produced per kg of fuel	m^3 / kg_{fuel}
n	Number of moles	mol
T	Average fuel gas temperature	K

P	Absolute pressure	Pa
C_{lim}	Exposure concentration limit	ppm
C_x	Actual Exposure concentration	ppm
EOR	Enhanced oil recovery	
SCR	Selective catalytic reduction	
OSHA	Occupational Safety and Health Agency	
TLV	Threshold limit value	
LD_{50}	50% Lethal dose	
ppm	Parts per million	

$NaHCO_3$	Sodium bi-carbonate	solid
Na_2CO_3	Sodium carbonate	solid
$CaCO_3$	Calcium carbonate	solid
$CaSO_3$	Calcium sulphite	solid
Na_2SO_4	Sodium sulphate	solid
NH_3	Ammonia	gas
CO_2	Carbon dioxide	gas
CO	Carbon monoxide	gas
H_2O	Water	gas
H_2S	Hydrogen Sulphide	gas
NO	Nitric oxide	gas
NO_2	Nitrogen dioxide	gas
SO_2	Sulphur dioxide	gas
N_2	Nitrogen gas	gas
O_2	Oxygen gas	gas
C	Carbon	
H	Elemental Hydrogen	
O	Elemental oxygen	
N	Elemental Nitrogen	
S	Elemental Sulphur	

CHAPTER 9: Summary of key findings and recommendations

The main purpose of this study was to investigate the feasibility of batch production of eco-fuel briquettes by studying processing parameters such as mixing characteristics, briquetting pressure, combustion characteristics and gas emissions analysis. A basic economic model was set up using the data from the study as input variables, in order to evaluate the economic feasibility of the proposed process.

Sourcing of the proposed raw material for the eco-fuel briquettes production has been the key element for the feasibility evaluation of the proposed process. The study has shown that the raw material used in making eco-fuel briquettes are traditionally waste material from other processes. This project helps to reduce waste disposal to the land fill sites, thus saving the waste generator landfill disposal costs, while simultaneously caring for the environment by saving the air space. However, a transport leg has been included in the economic model, for collection of raw material from the source to the production site. It could be negotiated so that the waste generators carry the transport costs, as they are already saving from landfill disposal gate fees.

The key finding obtained during the study of mixing characteristics of the raw materials was that the coal fines particle size in the mix affects the mixing index of the mixture, as well as the energy requirements for mixing. Furthermore, if coarser coal fines particles are used, it creates larger voids between the particles, resulting in poor binding strength and reduced briquettes stability. The coal fines must be screened such that 80% of the fines used in the mixture pass through a 500μm sieve screen. It was proposed that the screening of the coal fine could be done manually. When the correct particle size of coal fines are used, the mixture reaches uniformity quicker during mixing and less energy is consumed during mixing. The test results have shown that an average mixing energy requirement of 0.04kW.hr per kilogram of material to be mixed is required, this is evidence that mixing can also be conducted manually using a hand driven mixer.

The batch production of the eco-fuel briquettes was conducted using the Porta press, screw press and the Bikernmayer lab scale press. The samples were made from a sludge mixture containing: 31.91 % spent coffee beans, 23.04% finely grinded coal fines, 11.08% saw dust, 17.73% mielie husks, and 10.34% granulated paper and 5.91% pulp containing waste water. The above recipe may be varied depending on the availability of raw material. Based on the findings, the briquettes produced using above mentioned equipment were of good quality, however the screw press had the longest cycle time of 3.5 minutes compared to thc to the Porta press and Bikernmayer press with cycle times of 30 seconds and 10.2 seconds respectively.

A prototype Bikernmayer press is proposed, which is capable of producing four eco-fuel briquettes per cycle, making it four times greater than the Porta and screw presses in capacity. The fabrication cost of the Bikernmayer press is lower as compared to the other two presses. The operating pressures of the Porta, Bikernmayer and Screw presses are 0.11, 0.753 and 0.878 MPa respectively, and this is lower than the briquetting pressure stated in the literature 1.5- 3 MPa(Husain, Zainac & Abdullah, 2002). However, the pressures in literature were obtained using electrically operated hydraulic press, whereas a focus point in this project is to limit the dependency on electrical power. It has also been observed that the briquettes made from the screw press were more stable and compact, compared to the others with a density of 732 kg/m^3 and moisture content of 28.87%. The proposed prototype Bikernmayer press is thought to be best suited for batch production of eco-fuel briquettes due to its low cost, capacity and availability. The press has less moving parts, which makes it safer to be operated by two people. The test results have proven that the briquettes can be sufficiently compacted without applying a significant amount of pressure. This was confirmed by visual inspection of the briquettes produced. The Bikernmayer press can operate at a pressure of 0.753 MPa and the briquettes produced have a density of 645kg/m^3 and moisture content of 33.48%. Briquetting pressures for both Porta press and screw press were obtained by calculations; this could be more accurate if pressure measuring devices were installed in both equipments.

The drying characteristics of the fuel briquette in an automatic convective dryer were investigated. The drying pattern follows the one described by Perry & Green, 1998. Although the drying test was conducted for 9.5 hours before all the drying regimes may be identified, it was noticed that 35.67% moisture is driven off during the warm-up period, followed by 60.5% moisture reduction during the constant rate period and 3.83% during falling rate period. The warm-up and constant rate period takes 5.3 hour at this stage the briquette is dried to about 12.5% moisture. In the refractory brick fireplace, a drying heat requirement flux of 1,156 W/m^2 is expected, which includes the radiation component (66.5% of total). This implies that 1,156W will be required to dry 100 briquettes to 10.73% moisture within a period of 4.8 hours. The most economical method is to use the fuel briquette as a source of energy for the dying process, as it costs less than R1.06/kg to produce eco-fuel briquettes, with an energy value of 11.9c per MJ for a briquette, while the electricity price is 12.5c per MJ (before the 2010 increases). The average briquette surface temperature during drying is estimated to be 32 ^{0}C, which is reasonable for convection and radiation drying at atmospheric pressure. This will also prevent driving off the volatiles from the eco-fuel briquettes and possible ignition. Based on the combustion finding discussed in chapter 6 of this dissertation, eight dry briquettes each weighing 118grams at a net calorific value of 18.9 MJ/kg and heat transfer efficiency of 87.6%, will provide 17.58 MJ of heat energy. If this is energy is to be transferred over 4.6 hours drying time, it will yields 1,156W drying heat rate. In simple terms, eight eco-fuel briquettes may be used to dry 100 briquettes. The heat transferred by the briquettes can evaporate water at a rate of 0.285g/min. The proposed drying facility is similar to a fire place with refractory bricks on the inside and base floor to prevent heat loss to the surroundings.

It is recommended that more work is conducted to investigate the drying profile on direct heat drying. The wet briquettes could be dried on the steel grid placed perpendicular to the burning flames of the briquettes like a braai. Other wet briquettes can be stacked in a multi tray above the drying briquettes to improve energy efficiency by using the waste heat to pre dry the wet briquettes.

Combustion of eco-fuel briquettes in a laboratory-scale POCA ceramic stove was investigated to evaluate its combustion characteristics and gas emission quality using the Testo portable emission analyzer model 350s.The efficiencies were between 91-95% for carbon utilisation efficiency and over 99.47% for CO to CO_2

combustion efficiency at an estimated air-to-fuel ratio of 1.44. The average burning rate of 2g/min was obtained from the test work, meaning that 1kg of eco-fuel briquettes can burn for 7 hours which makes it ideal for domestic applications. The standard enthalpy of formation of the briquettes was estimated as $-1,619.3 kJmol^{-1}$ and the nett heat loss to the surroundings was $364.6Js^{-1}$ of which 10.7% was lost via free convection and balance by radiation. Some of these heat losses may be avoided by controlling the combustion conditions, for example pre-heating the air prior to combustion. 31.4% of the heat produced by the briquettes can be absorbed by a pot of water. The flue gas produced from this reaction consisted of 21.3% carbon dioxide, 0.0074% carbon monoxide, 0.000432% hydrogen sulphide, 0.000134% nitric oxide, 0.000367% sulphur dioxide, 5.5 % water vapour, 0.00598% hydrogen, 60.3 % nitrogen and 12.8 % oxygen. The gas was mostly dominated by inert atmospheric nitrogen, oxygen, carbon dioxide and water vapour. The test results have shown that an eco- fuel briquette is the cleanest solid fuel as compared to other fossil fuels composed primarily of 36.65% Carbon, 36.3% Oxygen, 4.6% Hydrogen, 0.75% Nitrogen, and 0.34% Sulphur and 10.46% Ash as per test results from ICP analysis.

It is recommended that more work is conducted aiming at determining the effects of operating parameters on eco-fuel briquettes combustion, such as gas velocity and excess air, preheated air, temperature and velocity may need to be investigated further in detail in order to improve the combustion efficiency of the briquettes under given conditions.

The current economic model is based at a selling price R2.26 per kilogram for the ease of accommodating the local market. This represents a lower cost per MJ than electricity. The financial model for both Porta press and screw press are not economically viable as their running costs were greater than the gross project revenue.

The Capital investment details as follows:

- Operating life time: 5years
- Direct Capital is R 106,873.44 + VAT
- Direct operating costs (also to be capitalised) at R563,108 + VAT,
- Total Capital investment is at R669,981+ VAT
- Gross Profit margin : 44%
- Profitability index of: 5.33
- Internal Rate of return :31.4%
- Net Present value: R676,896
- Return period: 0.408 years

The customer profile as currently at hand is 17% of the selected area within 80m radius from production site. The remaining 83% will be in need of energy as they become aware of the new product offering. The principal driver for this project is socio economic development, and it is being strengthened by Eskom's inability to provide sufficient energy. The secondary driver is the global drive to reduce emissions and fossil fuel usage, this technology does exactly this whilst diverting waste from landfill.

The production of eco-fuel briquettes is primarily targeting the local market, with the aim of creating a basic platform for socio-economic development by creating jobs and providing necessary skills to the communities. The project has environmental benefits, such as reducing the amount of general biomass waste going into the municipal landfills. As the product is considered a renewable energy source, the project has also carbon offset credits which may be earned in the form of revenue or tax credits. The health and safety of consumers is not compromised, as the briquettes burn cleanly within the acceptable emission exposure limits set out by OSHA. However, consumers need to be educated of the health hazards associated with the exposure of combustion gasses over prolonged periods, and without sufficient ventilation, in addition to the dangers related with the use of uncertified combustion appliances.

Most of heating appliances rely on the availability of air within the living space for the supply of combustion air. This includes unvented heaters, coal stoves, and other combustion appliances that also vent combustion products into the indoors. However, if the home is insulated and the air tightness increased, then these appliances are polluting the home and should be vented to the outside. Combustion devices should not only be fitted with a direct exhaust (chimney), but also with a dedicated intake, in order to supply the unit with combustion air. Ventilation is a necessary requirement for the proper operation of any home. Many existing homes rely on natural ventilation to control interior humidity, odours and combustion products. Natural ventilation is wind dependant, and therefore unreliable. The quality of construction, however, is improving, and homes are becoming tighter and more energy efficient. These houses cannot rely on natural ventilation, and as a result, new residential buildings are developing a need for mechanical ventilation.

Printed in Great Britain
by Amazon

48636542R00115